Heiko Dietrich

p-groups of maximal class

Heiko Dietrich

p-groups of maximal class

Periodic structures in the graph associated with p-groups of maximal class

Südwestdeutscher Verlag für Hochschulschriften

Impressum/Imprint (nur für Deutschland/ only for Germany)
Bibliografische Information der Deutschen Nationalbibliothek: Die Deutsche Nationalbibliothek verzeichnet diese Publikation in der Deutschen Nationalbibliografie; detaillierte bibliografische Daten sind im Internet über http://dnb.d-nb.de abrufbar.

Alle in diesem Buch genannten Marken und Produktnamen unterliegen warenzeichen-, marken- oder patentrechtlichem Schutz bzw. sind Warenzeichen oder eingetragene Warenzeichen der jeweiligen Inhaber. Die Wiedergabe von Marken, Produktnamen, Gebrauchsnamen, Handelsnamen, Warenbezeichnungen u.s.w. in diesem Werk berechtigt auch ohne besondere Kennzeichnung nicht zu der Annahme, dass solche Namen im Sinne der Warenzeichen- und Markenschutzgesetzgebung als frei zu betrachten wären und daher von jedermann benutzt werden dürften.

Verlag: Südwestdeutscher Verlag für Hochschulschriften Aktiengesellschaft & Co. KG
Dudweiler Landstr. 99, 66123 Saarbrücken, Deutschland
Telefon +49 681 37 20 271-1, Telefax +49 681 37 20 271-0, Email: info@svh-verlag.de
Zugl.: Braunschweig, Technische Universität Braunschweig, Dissertation, 2009

Herstellung in Deutschland:
Schaltungsdienst Lange o.H.G., Zehrensdorfer Str. 11, D-12277 Berlin
Books on Demand GmbH, Gutenbergring 53, D-22848 Norderstedt
Reha GmbH, Dudweiler Landstr. 99, D- 66123 Saarbrücken
ISBN: 978-3-8381-1005-9

Imprint (only for USA, GB)
Bibliographic information published by the Deutsche Nationalbibliothek: The Deutsche Nationalbibliothek lists this publication in the Deutsche Nationalbibliografie; detailed bibliographic data are available in the Internet at http://dnb.d-nb.de.
Any brand names and product names mentioned in this book are subject to trademark, brand or patent protection and are trademarks or registered trademarks of their respective holders. The use of brand names, product names, common names, trade names, product descriptions etc. even without
a particular marking in this works is in no way to be construed to mean that such names may be regarded as unrestricted in respect of trademark and brand protection legislation and could thus be used by anyone.

Publisher:
Südwestdeutscher Verlag für Hochschulschriften Aktiengesellschaft & Co. KG
Dudweiler Landstr. 99, 66123 Saarbrücken, Germany
Phone +49 681 37 20 271-1, Fax +49 681 37 20 271-0, Email: info@svh-verlag.de

Copyright © 2008 Südwestdeutscher Verlag für Hochschulschriften Aktiengesellschaft & Co. KG and licensors
All rights reserved. Saarbrücken 2008

Produced in USA and UK by:
Lightning Source Inc., 1246 Heil Quaker Blvd., La Vergne, TN 37086, USA
Lightning Source UK Ltd., Chapter House, Pitfield, Kiln Farm, Milton Keynes, MK11 3LW, GB
BookSurge, 7290 B. Investment Drive, North Charleston, SC 29418, USA
ISBN: 978-3-8381-1005-9

Dedicated to my grandfather

Hans Viktor Klein

Summary

A finite group whose order is a power of a prime p is called a finite p-group. Among finite groups, p-groups take a special position: For example, every finite group contains *large* p-groups as subgroups by the Sylow Theorems. The classification of p-groups is a difficult problem and, in general, not even the exact number of isomorphism types of groups of order p^n is known. The asymptotic estimates of Higman (1960) and Sims (1965) show that there are $p^{2n^3/27+O(n^{8/3})}$ isomorphism types of groups of order p^n. Higman's PORC Conjecture (1960) claims that for fixed n the number of isomorphism types of groups of order p^n is a polynomial on residue classes.

A special type of p-groups are p-groups of maximal class: These are the p-groups of order p^n with nilpotency class $n-1$. A first major study of maximal class groups was carried out by Blackburn in 1958. He obtained a classification of the 2- and 3-groups of maximal class, and he observed that a classification for primes greater than 3 is significantly more difficult. Following Blackburn, maximal class groups were discussed in detail by Shepherd (1971), Miech (1970 – 1982), Leedham-Green & McKay (1976 – 1984), Fernández-Alcober (1995), and Vera-López et al. (1995 – 2008). Despite substantial progress made in the last six decades, the classification of maximal class groups is still an open problem in p-group theory. For example, Problem 3 of Shalev's survey paper (1994) on finite p-groups asks to classify the 5-groups of maximal class.

The coclass of a p-group of order p^n and nilpotency class c is defined as $n-c$. Hence, the p-groups of maximal class are the p-groups of coclass 1. Leedham-Green & Newman (1980) suggested to classify p-groups by coclass, and their suggestion has led to a major research project in p-group theory. In this thesis, we follow the philosophy of coclass theory and investigate the p-groups of maximal class (or coclass 1).

The graph $\mathcal{G}(p)$. The coclass graph $\mathcal{G}(p)$ associated with p-groups of maximal class is defined as follows: Its vertices are the isomorphism types of finite p-groups of maximal class where a vertex is identified with a group representing its isomorphism class. Two vertices G and H are connected by a directed edge $G \to H$ if and only if G is isomorphic to the central quotient $H/\zeta(H)$. A group H is called a descendant of a group G in $\mathcal{G}(p)$ if $G = H$ or if there is a path from G to H. We visualize $\mathcal{G}(p)$ in the Euclidean plane as an undirected graph by drawing the proper descendants of a group in $\mathcal{G}(p)$ below that group.

It is well-known that $\mathcal{G}(p)$ consists of the cyclic group of order p^2 and an infinite tree $\mathcal{T}(p)$, whose root is elementary abelian of order p^2. The tree $\mathcal{T}(p)$ contains a unique infinite path starting at its root. This path is called the mainline of $\mathcal{T}(p)$ and we denote it by $S_2 \to S_3 \to \ldots$ where S_n has order p^n. The n-th branch $\mathcal{B}_n$ of the tree $\mathcal{T}(p)$ is the finite subtree of $\mathcal{T}(p)$ induced by the descendants of S_n which are not descendants of S_{n+1}. As usual, the depth of a subtree $\mathcal{B}$ of $\mathcal{T}(p)$ is the maximal length of a path within $\mathcal{B}$, and its width is the maximal number of vertices at the same depth in $\mathcal{B}$. A sketch of $\mathcal{G}(p)$ is given in Figure 1.

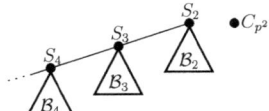

Figure 1: The graph $\mathcal{G}(p)$.

By the results of Blackburn, the graph $\mathcal{G}(p)$ for $p = 2, 3$ is completely understood. For example, all branches have depth 1 and a recurrent structure. Leedham-Green & McKay (1984) showed that the set of depths of the branches of $\mathcal{T}(p)$ is unbounded for $p \geq 5$ and the set of widths of the branches of $\mathcal{T}(5)$ is bounded. Here we prove that the set of widths of the branches of $\mathcal{T}(p)$ is unbounded for $p \geq 7$. For $p = 5$ an investigation with computational methods is still possible, see Newman (1990) and Dietrich, Eick & Feichtenschlager (2008), whereas for $p \geq 7$ the size of the branches increases too fast for a complete examination. The detailed structure of $\mathcal{G}(p)$ for $p \geq 7$ is not known.

The periodicity of type 1. We now describe our first main results and, for this purpose, we introduce some more notation. For $k \geq 0$ the shaved branch $\mathcal{B}_n[k]$ is the subtree of $\mathcal{B}_n$ induced by the groups of distance at most k from its root S_n. We define functions $\mathfrak{c} = \mathfrak{c}(p)$ and $\mathfrak{e}_n = \mathfrak{e}_n(p)$, both essentially given by linear polynomials, which for given p satisfy $0 \leq \mathfrak{e}_2 \leq \mathfrak{e}_3 \leq \ldots$ and $\mathfrak{e}_{n+d} = \mathfrak{e}_n + d$ where $d = p - 1$. We call the shaved branch $\mathcal{T}_n = \mathcal{B}_n[\mathfrak{e}_n]$ the n-th body of $\mathcal{T}(p)$ and prove the following for $n \geq p + 1$.

- The depths of $\mathcal{B}_n$ and $\mathcal{T}_n$ differ by at most $\mathfrak{c}$.
- There is an embedding $\iota = \iota_n \colon \mathcal{T}_n \hookrightarrow \mathcal{B}_{n+d}$ of rooted trees such that $\iota(\mathcal{T}_n) = \mathcal{B}_{n+d}[\mathfrak{e}_n]$.

A summary of these results is visualized in Figure 2.

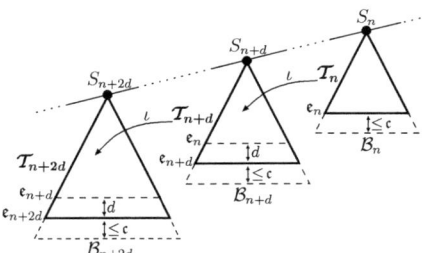

Figure 2: The periodicity of type 1.

This recurrent pattern in $\mathcal{T}(p)$ is referred to as the periodicity of type 1. It shows that a major part of the tree carries a periodic structure. Weaker versions of this periodic pattern in $\mathcal{T}(p)$ have been proved by Eick & Leedham-Green (2008) and du Sautoy (2001).

Experiments by computer suggest that the whole branch $\mathcal{B}_n$ cannot be embedded into $\mathcal{B}_{n+d}$. The periodicity of type 1 as proved in this thesis embeds $\mathcal{B}_n$ with the exclusion of at most $\mathfrak{c}$ levels of groups. This shows that the periodicity of type 1 is close to best possible.

The periodicity of type 2. For a complete description of the tree $\mathcal{T}(p)$ it is necessary to describe the difference graph of $\mathcal{B}_{n+d}$ and $\iota(\mathcal{T}_n) = \mathcal{B}_{n+d}[\mathfrak{e}_n]$. Since in general the set of widths of the branches is unbounded, this graph cannot be isomorphic to a subgraph of $\mathcal{B}_n$. However, a conjecture of Eick, Leedham-Green, Newman & O'Brien (2009) claims that it can be described by another periodic pattern, that is, a periodicity of type 2. According to this conjecture, for large enough n, the subtree of $\mathcal{B}_{n+d}$ induced by the descendants of a group at depth $\mathfrak{e}_n$ in $\mathcal{B}_{n+d}$ is isomorphic to a corresponding subtree in $\mathcal{B}_n$. Thus, the periodicities of type 1 and 2 would in principle suffice to describe the tree $\mathcal{T}(p)$ completely.

As a first approximation of this conjecture, we consider the difference graph of $\mathcal{T}_{n+d}$ and $\iota(\mathcal{T}_n)$ and, thus, omit at most $\mathfrak{c}$ levels of groups. We define the d-step descendant tree $\mathcal{D}_d(G)$ of a group G in $\mathcal{T}(p)$ as the subtree of $\mathcal{T}(p)$ induced by the descendants of distance at most d from G. Then the periodicity of type 2 asserts that for large enough n every group G at depth $\mathfrak{e}_n$ in $\mathcal{T}_{n+d}$ has a periodic parent H at depth $\mathfrak{e}_n - d$ in $\mathcal{T}_{n+d}$ such that $\mathcal{D}_d(H)$ and $\mathcal{D}_d(G)$ are isomorphic as rooted trees, see Figure 3.

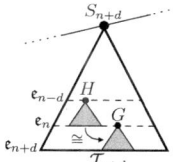

Figure 3: A periodic parent of G.

This periodic pattern and the periodicity of type 1 would suffice to describe the bodies of $\mathcal{T}(p)$ completely. The main problem is to specify a periodic parent of a group. Based on ideas of Leedham-Green & McKay (1984), we use p-adic number theory and prove the periodicity of type 2 in certain special cases. There are significant differences depending on the residue p modulo 6, and we consider the easier case $p \equiv 5 \bmod 6$ here. As a corollary, we show that for large enough n the d-step parent of a group at depth $\mathfrak{e}_n$ in $\mathcal{T}_{n+d}$ which has proper descendants is a periodic parent if its automorphism group is a p-group. The computation of explicit examples indicate that the d-step parent is not always a periodic parent, and we propose an alternative construction of periodic parents in a special case.

Classification of groups. The periodicities of type 1 and 2 describe graph theoretic patterns within $\mathcal{T}(p)$. For our proof of these periodicities we use a cohomological approach and describe the groups in the bodies of $\mathcal{T}(p)$ as certain group extensions. This allows us to construct all graph isomorphisms on a group theoretic level such that the periodic patterns in $\mathcal{T}(p)$ are reflected in the structure of the groups involved. For example, if G is a group in $\mathcal{T}_n$ with $n \geq p+1$, then a suitable choice of the embeddings ι allows us to describe the infinitely many groups in $\{G, \iota(G), \iota^2(G), \ldots\}$ by a single group presentation whose defining relations contain one indeterminate integer as parameter.

The 5-groups of maximal class. As an application, we prove that the bodies $\mathcal{T}_n = \mathcal{B}_n[n-4]$ of the tree $\mathcal{T}(5)$ can be described by a finite subgraph and the periodicities of type 1 and 2. We deduce that the infinitely many groups in these bodies can be described by finitely many group presentations with at most two indeterminate integers as parameters. This is close to a positive answer of Problem 3 in Shalev's survey paper (1994).

Zusammenfassung

Eine endliche p-Gruppe ist eine Gruppe mit Primzahlpotenzordnung p^n. In der Klasse der endlichen Gruppen nehmen p-Gruppen eine besondere Stellung ein: Beispielsweise folgt aus den Sylow-Sätzen, dass jede endliche Gruppe *große* p-Gruppen als Untergruppen enthält. Die Klassifikation von p-Gruppen ist ein schwieriges Unterfangen und im Allgemeinen ist nicht einmal die exakte Anzahl der Isomorphietypen von Gruppen der Ordnung p^n bekannt. Asymptotische Abschätzungen von Higman (1960) und Sims (1965) zeigen, dass es $p^{2n^3/27+O(n^{8/3})}$ Isomorphietypen von Gruppen der Ordnung p^n gibt. Higmans PORC Vermutung (1960) sagt voraus, dass die Anzahl der Isomorphietypen von Gruppen der Ordnung p^n für festes n ein Polynom auf Restklassen ist.

Ein besonderer Typ von p-Gruppen sind die p-Gruppen mit maximaler Klasse. Dies sind die p-Gruppen der Ordnung p^n mit Nilpotenzklasse $n-1$. Eine erste grundlegende Untersuchung von Gruppen mit maximaler Klasse wurde 1958 von Blackburn vorgenommen. Blackburn erzielte die Klassifikation der 2- und 3-Gruppen mit maximaler Klasse und er beobachtete, dass eine Klassifikation für größere Primzahlen weitaus schwieriger ist. Weitere detaillierte Untersuchungen der Gruppen mit maximaler Klasse wurden von Shepherd (1971), Miech (1970 – 1982), Leedham-Green & McKay (1976 – 1984), Fernández-Alcober (1995) und Vera-López et al. (1995 – 2008) durchgeführt. Trotz erheblichen Fortschritts in den letzten sechs Jahrzehnten ist die Klassifikation der Gruppen mit maximaler Klasse noch immer ein offenes Problem in der Theorie der p-Gruppen. Beispielsweise fragt Shalev (1994) in Problem 3 seines Übersichtsartikels über p-Gruppen nach einer Klassifikation der 5-Gruppen mit maximaler Klasse.

Die Koklasse einer p-Gruppe mit Ordnung p^n und Nilpotenzklasse c ist definiert als $n-c$, das heißt, p-Gruppen mit maximaler Klasse entsprechen den p-Gruppen mit Koklasse 1. Leedham-Green & Newman (1980) machten den Vorschlag, p-Gruppen nach ihrer Koklasse zu klassifizieren, und legten damit den Grundstein für ein umfangreiches Forschungsprojekt in der Theorie der p-Gruppen. In der vorliegenden Arbeit wird diese Theorie benutzt, um die p-Gruppen mit maximaler Klasse (oder Koklasse 1) zu untersuchen.

Der Graph $\mathcal{G}(p)$. Der den p-Gruppen mit maximaler Klasse zugeordnete Koklassengraph $\mathcal{G}(p)$ ist wie folgt definiert: Die Knoten sind die Isomorphietypen von endlichen p-Gruppen mit maximaler Klasse, wobei ein Knoten mit einem Isomorphietyp-Repräsentanten identifiziert wird. Zwei Knoten G und H sind genau dann mit einer gerichteten Kante $G \to H$ verbunden, wenn G isomorph zu dem zentralen Quotienten $H/\zeta(H)$ ist. Eine Gruppe H heißt Nachfolger einer Gruppe G in $\mathcal{G}(p)$, falls $G = H$ oder falls es einen Pfad von G nach H gibt. Der gerichtete Graph $\mathcal{G}(p)$ wird in der Euklidischen Ebene ungerichtet dargestellt, indem die echten Nachfolger einer Gruppe in $\mathcal{G}(p)$ unterhalb dieser Gruppe gezeichnet werden. Es ist bekannt, dass sich $\mathcal{G}(p)$ aus der zyklischen Gruppe der Ordnung p^2 und einem unendlichen Baum $\mathcal{T}(p)$ mit elementar-abelscher Wurzel der Ordnung p^2 zusammensetzt. Die Wurzel von $\mathcal{T}(p)$ ist der Startknoten eines eindeutigen unendlichen Pfades. Dieser Pfad ist die Hauptlinie von $\mathcal{T}(p)$ und wird als $S_2 \to S_3 \to \ldots$ bezeichnet, wobei S_n die Ordnung p^n hat. Der n-te

Ast $\mathcal{B}_n$ des Baumes $\mathcal{T}(p)$ ist der endliche Teilbaum von $\mathcal{T}(p)$, der von den Nachfolgern von S_n induziert wird, welche nicht auch Nachfolger von S_{n+1} sind. Wie üblich sind Tiefe und Weite eines Wurzelbaumes definiert als die maximale Länge eines Pfades, beziehungsweise die maximale Anzahl von Knoten der gleichen Tiefe. In Abbildung 1 ist die Struktur von $\mathcal{G}(p)$ skizziert.

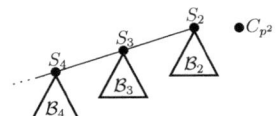

Abbildung 1: Der Graph $\mathcal{G}(p)$.

Die Resultate von Blackburn beschreiben die Graphen $\mathcal{G}(p)$ für $p = 2, 3$ vollständig und es ist bekannt, dass alle Äste die Tiefe 1 und periodisch auftretende Struktur haben. Leedham-Green & McKay (1984) haben gezeigt, dass für $p \geq 5$ die Menge der Tiefen der Äste von $\mathcal{T}(p)$ unbeschränkt und die Menge der Weiten der Äste von $\mathcal{T}(5)$ beschränkt ist. In der vorliegenden Arbeit wird bewiesen, dass die Menge der Weiten der Äste von $\mathcal{T}(p)$ unbeschränkt ist für $p \geq 7$. Während $\mathcal{T}(5)$ daher detailliert mit dem Computer untersucht werden kann, siehe zum Beispiel Newman (1990) und Dietrich, Eick & Feichtenschlager (2008), so ist eine solche ausführliche Untersuchung für $p \geq 7$ auf Grund des Wachstums der Äste nicht möglich. Die detaillierte Struktur von $\mathcal{G}(p)$ für $p \geq 7$ ist daher unbekannt.

Die Periodizität vom Typ 1. Um das erste Hauptresultat dieser Arbeit zu beschreiben, ist weitere Notation nötig. Für $k \geq 0$ ist der gestutzte Ast $\mathcal{B}_n[k]$ der Teilbaum von $\mathcal{B}_n$, welcher von den Gruppen der Tiefe höchstens k in $\mathcal{B}_n$ induziert wird. Weiterhin werden Abbildungen, im Wesentlichen lineare Polynome, $\mathfrak{c} = \mathfrak{c}(p)$ und $\mathfrak{e}_n = \mathfrak{e}_n(p)$ definiert mit $0 \leq \mathfrak{e}_2 \leq \mathfrak{e}_3 \leq \ldots$ und $\mathfrak{e}_{n+d} = \mathfrak{e}_n + d$ für festes p und $d = p - 1$. Der n-te Rumpf von $\mathcal{T}(p)$ ist der gestutzte Ast $\mathcal{T}_n = \mathcal{B}_n[\mathfrak{e}_n]$. Folgende Aussagen werden für $n \geq p + 1$ bewiesen:

- Die Tiefen von $\mathcal{B}_n$ und $\mathcal{T}_n$ unterscheiden sich höchstens um $\mathfrak{c}$.
- Es gibt eine Einbettung $\iota = \iota_n \colon \mathcal{T}_n \hookrightarrow \mathcal{B}_{n+d}$ von Wurzelbäumen mit $\iota(\mathcal{T}_n) = \mathcal{B}_{n+d}[\mathfrak{e}_n]$.

Eine Zusammenfassung dieser Ergebnisse ist in Abbildung 2 skizziert.

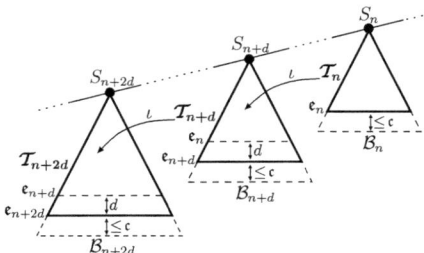

Abbildung 2: Die Periodizität vom Typ 1.

Dieses periodisch auftretende Muster in $\mathcal{T}(p)$ wird als Periodizität vom Typ 1 bezeichnet. Es zeigt, dass ein Großteil des Baumes eine periodische Struktur trägt. Schwächere Versionen dieser Periodizität wurden von Eick & Leedham-Green (2008) und du Sautoy (2001) bewiesen. Explizite Berechnungen mit dem Computer lassen vermuten, dass man nicht den gesamten Ast $\mathcal{B}_n$ in $\mathcal{B}_{n+d}$ einbetten kann. Die in dieser Arbeit bewiesene Periodizität vom Typ 1 bettet $\mathcal{B}_n$ mit Ausnahme von höchstens $\mathfrak{c}$ Levels von Gruppen in den Ast $\mathcal{B}_{n+d}$ ein. Dies deutet darauf hin, dass die Periodizität vom Typ 1 nahezu bestmöglich ist.

Die Periodizität vom Typ 2. Für eine vollständige Beschreibung des Baumes $\mathcal{T}(p)$ muss der Differenzengraph von $\mathcal{B}_{n+d}$ und $\iota(\mathcal{T}_n) = \mathcal{B}_{n+d}[\mathfrak{e}_n]$ beschrieben werden. Im Allgemeinen ist die Menge der Weiten der Äste unbeschränkt, weswegen dieser Differenzengraph nicht zu einem Teilgraphen von $\mathcal{B}_n$ isomorph sein kann. Eine Vermutung von Eick, Leedham-Green, Newman & O'Brien (2009) besagt jedoch, dass er mit Hilfe eines weiteren periodischen Musters, einer Periodizität vom Typ 2, beschrieben werden kann. Entsprechend dieser Vermutung ist für genügend großes n der Teilbaum von $\mathcal{B}_{n+d}$, der von den Nachfolgern einer Gruppe der Tiefe $\mathfrak{e}_n$ in $\mathcal{B}_{n+d}$ induziert wird, isomorph zu einem entsprechenden Teilbaum in $\mathcal{B}_n$. Die Periodizitäten vom Typ 1 und 2 wären danach ausreichend, um die Äste in $\mathcal{T}(p)$ vollständig zu beschreiben.

Als eine erste Annäherung an diese Vermutung wird in der vorliegenden Arbeit der Differenzengraph von $\mathcal{T}_{n+d}$ und $\iota(\mathcal{T}_n)$ betrachtet, das heißt, es werden höchstens $\mathfrak{c}$ Levels von Gruppen ausgelassen. Der Nachfolgerbaum $\mathcal{D}_d(G)$ einer Gruppe G in $\mathcal{T}(p)$ ist der Teilbaum von $\mathcal{T}(p)$, welcher von den Nachfolgern mit Abstand höchstens d von G induziert wird. Die Periodizität vom Typ 2 sagt nun voraus, dass für genügend großes n jede Gruppe G der Tiefe $\mathfrak{e}_n$ in $\mathcal{T}_{n+d}$ einen periodischen Vorfahren H der Tiefe $\mathfrak{e}_n - d$ in $\mathcal{T}_{n+d}$ besitzt, so dass die Nachfolgerbäume $\mathcal{D}_d(H)$ und $\mathcal{D}_d(G)$ isomorph sind, siehe Abbildung 3.

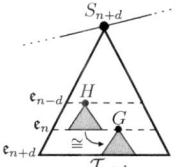

Abbildung 3: Ein periodischer Vorfahre von G.

Dieses periodische Muster und die Periodizität vom Typ 1 wären ausreichend, um die Rümpfe von $\mathcal{T}(p)$ vollständig zu beschreiben. Das Hauptproblem ist, einen periodischen Vorfahren einer Gruppe zu bestimmen. In der vorliegenden Arbeit wird, basierend auf Ideen von Leedham-Green & McKay (1984), p-adische Zahlentheorie benutzt, um die Periodizität vom Typ 2 in gewissen Spezialfällen zu beweisen. Dabei ergeben sich signifikante Unterschiede zwischen den verschiedenen Restklassen p modulo 6, so dass hier lediglich der einfachere Fall $p \equiv 5 \mod 6$ genauer untersucht wird. Als eine Folgerung wird gezeigt, dass der d-fache Vorfahre einer Gruppe der Tiefe $\mathfrak{e}_n$ in $\mathcal{T}_{n+d}$ mit echten Nachfolgern ein periodischer Vorfahre ist, falls seine Automorphismengruppe eine p-Gruppe ist. Die Untersuchung expliziter Beispiele deutet darauf hin, dass der d-fache Vorfahre nicht immer als periodischer Vorfahre gewählt werden kann. Eine alternative Konstruktion von periodischen Vorfahren wird in einem Spezialfall beschrieben.

Die Klassifikation von Gruppen. Die Periodizitäten vom Typ 1 und 2 beschreiben graphentheoretische Muster innerhalb des Baumes $\mathcal{T}(p)$. Für den hier geführten Beweis dieser Periodizitäten wird ein kohomologischer Ansatz verfolgt und die Gruppen in den Rümpfen von $\mathcal{T}(p)$ werden als gewisse Gruppenerweiterungen beschrieben. Hierdurch ist es möglich, die entsprechenden Graphenisomorphismen auf gruppentheoretischer Ebene zu konstruieren. Die periodischen Strukturen in $\mathcal{T}(p)$ spiegeln sich dabei in der Struktur der entsprechenden Gruppen wider. Ist G beispielsweise eine Gruppe in dem Rumpf $\mathcal{T}_n$ mit $n \geq p + 1$, dann erlaubt eine geeignete Wahl der Einbettungen ι eine Beschreibung der unendlich vielen Gruppen in $\{G, \iota(G), \iota^2(G), \ldots\}$ durch eine einzige Gruppenpräsentation deren definierenden Relatoren eine natürliche Zahl als Parameter besitzen.

Die 5-Gruppen mit maximaler Klasse. Als eine Anwendung der erzielten Ergebnisse wird gezeigt, dass die Rümpfe $\mathcal{T}_n = \mathcal{B}_n[n-4]$ des Baumes $\mathcal{T}(5)$ durch einen endlichen Teilgraphen und die Periodizitäten vom Typ 1 und 2 beschrieben werden können. Es wird gefolgert, dass die unendlich vielen Gruppen in diesen Rümpfen durch endlich viele Gruppenpräsentationen mit höchstens zwei natürlichen Zahlen als Parameter beschrieben werden. Dies kommt einer Beantwortung von Problem 3 in Shalevs Übersichtsartikel (1994) nahe.

Contents

Dedication	i
Summary	iii
Zusammenfassung	vii

1 Introduction — 1
  1.1 Classification by coclass 1
  1.2 Groups of maximal class 6
  1.3 Comment on the notation 7

2 The graph $\mathcal{G}(p)$ — 9
  2.1 General notation 9
  2.2 Bounding the depths 10
  2.3 Periodicity of type 1 11
  2.4 Periodicity of type 2 13
  2.5 The graph $\mathcal{G}(5)$ 15
  2.6 Open problems 15

3 Basics of maximal class groups — 17
  3.1 Normal subgroup structure 17
  3.2 Degree of commutativity 18
  3.3 Maximal class and uniserial action 19

4 Polycyclic groups and cohomology — 21
  4.1 Polycyclic groups 21
  4.2 Cohomology of polycyclic groups 24

5 Number theory — 31
  5.1 The p-th local cyclotomic field 31
  5.2 Pro-p groups and $\mathbb{Z}_p$-modules 34
  5.3 Homomorphisms from $T \wedge T$ 35
  5.4 The action of p-adic units 38

6 The space group of maximal class — 45
  6.1 Basic definitions 45
  6.2 Connection with the graph $\mathcal{G}(p)$ 47
  6.3 Standard presentations 49
  6.4 Automorphism groups 51
  6.5 Cohomology 53

7	**Cohomological description of maximal class groups**	**61**
	7.1 Tail vectors defining maximal class groups	61
	7.2 Isomorphism problem	64
	7.3 Twigs, skeleton, and capable groups	67
8	**Periodicity of type I**	**71**
	8.1 Graph isomorphisms	71
	8.2 Periodicity classes	73
9	**Periodicity of type 2**	**75**
	9.1 Descendant trees	76
	9.2 The action of the stabilizer	77
	9.3 The case $p \equiv 5 \bmod 6$	80
10	**5-groups of maximal class**	**85**
	10.1 The graph $\mathcal{G}(5)$	85
	10.2 Periodicity classes	88
A	**Appendix**	**91**
	A.1 Technical details	91
	A.2 Coclass conjectures	95

Bibliography   99

List of symbols   103

Index   107

Acknowledgments   109

1 Introduction

> *"The chief forms of beauty are order and symmetry and definiteness, which the mathematical sciences demonstrate in a special degree."*
>
> Aristotle (384 B.C. – 322 B.C.)

The concept of a group is a central concept of abstract algebra. Basically describing symmetries of certain objects, groups can be found in most branches of mathematics and also in many areas of science, for example physics, chemistry, coding theory, and cryptography.

One of the major themes in finite group theory is the classification of groups. The general idea of classification is to find for a given class of groups an explicit list of isomorphism type representatives; that is, no two groups in the list are isomorphic and every group in the given class is isomorphic to a group in the list. An example of a famous classification theorem is the *classification theorem of finite simple groups*, which classifies all finite simple groups – the basic building blocks of all finite groups. However, if we even restrict attention to the least complicated finite simple group, the cyclic group of prime order p, it is still an intricate problem to put these groups together in order to construct all groups of p-power order, so-called finite p-groups, up to isomorphism.

An approach to classify finite p-groups is to classify all groups of a given order, p^n say. Higman's PORC Conjecture (***p**olynomial **o**n **r**esidue **c**lasses*) claims that for fixed n there is an integer m such that the number of (isomorphism types of) groups of order p^n is a polynomial in p which depends on the residue class p modulo m, see [21]. The asymptotic estimates of Higman [20] and Sims [53] show that there are $p^{2n^3/27 + O(n^{8/3})}$ groups of order p^n. Thus, the number of groups grows exponentially with n and, apart from the same order, the groups of order p^n have the most different structures which complicates a possible classification. Newman, O'Brien & Vaughan-Lee [43, 45] classified the groups of order dividing p^7, and these groups are implemented in several computer algebra systems. For larger exponents n it seems hopeless to find such a description. For example, there is neither a theoretical classification nor an implementation of the 49487365422 groups of order $2^{10} = 1024$. This indicates that it might be useful to classify p-groups by other invariants than the order.

1.1 Classification by coclass

The original roots of *coclass theory* go back to the 1950s when groups of maximal class were examined, that is, finite p-groups of the largest nilpotency class compatible with their order. By definition, a finite p-group G of order p^n has maximal class if it has nilpotency class $c(G) = n - 1$. Thus, the lower central series of G has the form

$$G = \gamma_1(G) > \ldots > \gamma_n(G) = \{1\}.$$

Maximal class groups were introduced by Wiman [67] in 1952. His partially incorrect paper was to some extent the basis of the investigations of Blackburn [3] in 1958, which nowadays is seen as the first major study of maximal class groups. In contrast to the p-groups of a given order, there are infinitely many p-groups of maximal class. However, the possible structures of these groups are restricted which perhaps allows a description by a finite set of data. As an example, we now briefly recall the classical results about 2- and 3-groups of maximal class.

The 2-groups of maximal class have been determined in the beginning of the twentieth century, see for example Séguier [48, p. 121] or Taussky [55], cf. [23, Satz III.11.9]. It is well-known that they consist of the cyclic group and elementary abelian group of order 4, the quaternion and dihedral group of order 8, and the following three infinite families of dihedral, semi-dihedral, and quaternion groups of 2-power order. We describe these families by three "parameterized presentations", that is, group presentations whose defining relations have exponents which are arithmetic expressions containing finitely many indeterminate integers as parameters.

$$
\begin{aligned}
D_{2^n} &= \langle a, b \,|\, a^{2^{n-1}} = 1,\ b^2 = 1,\ a^b = a^{-1} \rangle & (n \geq 4), \\
SD_{2^n} &= \langle a, b \,|\, a^{2^{n-1}} = 1,\ b^2 = 1,\ a^b = a^{2^{n-2}-1} \rangle & (n \geq 4), \\
Q_{2^n} &= \langle a, b \,|\, a^{2^{n-1}} = 1,\ b^2 = a^{2^{n-2}},\ a^b = a^{-1} \rangle & (n \geq 4).
\end{aligned}
$$
(1.1)

In a similar way, Blackburn [3] determined group presentations of the 3-groups of maximal class. We modify these presentations and, without a proof, describe the groups as follows. Let $m \geq 3$. If $n = 2m$ is even, then there are seven maximal class groups of order 3^n which are defined by

$$
\begin{aligned}
P_{u,v,w} = \langle a,b,c,d \,|\, & a^3 = d^u,\ b^{3^{m-1}} = d,\ c^{3^{m-1}} = d,\ d^3 = 1, \\
& b^a = c,\ c^a = b^{3^{m-1}-1}c^{3^{m-1}-1}d^v,\ c^b = cd^w,\ d\ \text{central}\rangle
\end{aligned}
$$

with $(u,v,w) \in \{(0,0,0), (0,0,1), (0,0,2), (0,1,0), (0,2,0), (1,1,0), (1,1,1)\}$. If $n = 2m-1$ is odd, then there are six maximal class groups of order 3^n which are defined by

$$
\begin{aligned}
P_{u,v,w} = \langle a,b,c,d \,|\, & a^3 = d^u,\ b^{3^{m-1}} = c^{3^{m-1}}d,\ c^{3^m} = 1,\ d^3 = 1, \\
& b^a = c,\ c^a = b^{3^{m-1}-1}c^{2 \cdot 3^{m-1}-1}d^v,\ c^b = cd^w,\ d\ \text{central}\rangle
\end{aligned}
$$

with $(u,v,w) \in \{(0,0,0), (0,0,1), (0,0,2), (0,2,0), (1,0,1), (1,2,0)\}$.

1.1.1 Coclass

In 1980, inspired by Blackburn's approach to the groups of maximal class, Leedham-Green & Newman [30] started the more general *classification by coclass* project. They considered the coclass of a p-group as an invariant and suggested to classify by coclass.

Definition. A group G of order p^n and nilpotency class c has coclass $\mathrm{cc}(G) = n - c$.

Together with this definition, Leedham-Green & Newman proposed five conjectures, known as Conjectures A – E, on the structure of the p-groups of a fixed coclass, see Section A.2. Their suggestion and the related conjectures initiated a deep research project. An important milestone in this project is the complete proof of all five coclass conjectures. Various results have been obtained along the way, until finally two independent proofs emerged, see Shalev [50] and Leedham-Green [32]. A full account of the proofs including further details and references is given in the book of Leedham-Green & McKay [29].

1.1. Classification by coclass

Coclass theory has delivered significant insight into the structure of the p-groups of a fixed coclass. For example, Conjecture A, which now is a theorem, states that there exists a function $f(p, r)$ such that every finite p-group of coclass r has a normal subgroup of nilpotency class at most 2 and index at most $f(p, r)$, see Section A.2.

The coclass project now continues to investigate the situation in more detail. An important conjecture underpinning this research project is the following, see for example [6].

1.1 Conjecture. *Let p be a prime and let r be a positive integer. The finite p-groups of coclass r can be divided into finitely many periodicity classes such that the structure of the groups in a periodicity class can be described in a uniform way. In particular:*

a) *All groups in a periodicity class can be defined by a single parameterized presentation.*

b) *Many structural invariants of the groups in a periodicity class can be exhibited in a uniform way. For example, their Schur multiplicators and automorphism groups can be described by a single parameterized presentation.*

Conjecture 1.1 is rather vague as no precise definition of *periodicity class* is given. A concrete definition of this term would already be a major step forward in this research area and perhaps also a significant step ahead in proving the conjecture. Nonetheless, one can observe that if this conjecture is true, then a classification of p-groups by coclass is possible and would be a powerful tool in the understanding of p-groups.

In Section 1.1.3 we report on some evidence supporting Conjecture 1.1. As a first step, we define coclass graphs – an important tool in coclass theory.

1.1.2 Coclass graphs

The coclass graph $\mathcal{G}(p, r)$ associated with p-groups of coclass r is defined as follows. Recall that $c(H)$ denotes the nilpotency class of a nilpotent group H.

Definition. The vertices of $\mathcal{G}(p, r)$ are the isomorphism types of finite p-groups of coclass r where a vertex is identified with a group representing its isomorphism class. Two vertices G and H are connected by a directed edge $G \to H$ if and only if $G \cong H/\gamma_{c(H)}(H)$.

A group H in $\mathcal{G}(p, r)$ is a descendant of a group G if $G = H$ or if there is a path from G to H in $\mathcal{G}(p, r)$. If this path has length 1, then H is an immediate descendant of G and G is the parent of H. By definition, a coclass graph is a directed graph. However, we visualize it in the Euclidean plane as an undirected graph by drawing the proper descendants of a group in $\mathcal{G}(p, r)$ below that group.

Example. Well-understood examples are the coclass graphs $\mathcal{G}(2, 1)$ and $\mathcal{G}(3, 1)$, which are sketched in Figure 1.1. For $\mathcal{G}(3, 1)$ we use a more compact notation: A vertex labeled with an integer m stands for m immediate descendants of the corresponding parent.

This example is in some sense misleading as in general the structure of a coclass graph is much more complicated. However, it is known that its structure cannot be arbitrarily wild. The following deep result was obtained in the course of proving the coclass conjectures. We refer to the book of Leedham-Green & McKay [29] and the book of Dixon, du Sautoy, Mann & Segal [7, Chapter 10] for references and background, cf. Section A.2.

Theorem. *Excluding finitely many groups, the graph $\mathcal{G}(p, r)$ is a finite collection of infinite trees such that every tree has exactly one infinite path starting at its root.*

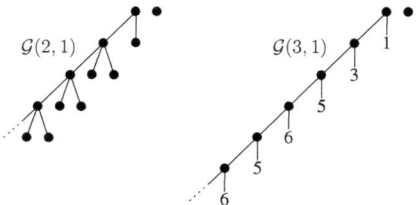

Figure 1.1: The coclass graphs $\mathcal{G}(2,1)$ and $\mathcal{G}(3,1)$.

A subtree of $\mathcal{G}(p,r)$ which is induced by all descendants of a group in $\mathcal{G}(p,r)$ is called a descendant tree. A coclass tree of $\mathcal{G}(p,r)$ is a descendant tree which is maximal with the property that its root is the initial vertex of exactly one infinite path. This unique maximal path is the mainline of the coclass tree. Using this notation, the graph $\mathcal{G}(p,r)$ consists of finitely many coclass trees and finitely many groups not lying in any coclass tree.

Closely connected to coclass trees are infinite pro-p groups, that is, infinite inverse limits of (topologically discrete) finite p-groups. The coclass of an infinite pro-p group S with nilpotent quotients $S_j = S/\gamma_j(S)$ for $j \geq 2$ is defined as $\mathrm{cc}(S) = r$ if there is an integer t such that S_j is a finite p-group of coclass r for all $j \geq t$. Obviously, such a pro-p group defines an infinite path $S_t \to S_{t+1} \to \ldots$ in the coclass graph $\mathcal{G}(p,r)$. On the other hand, the inverse limit of the groups on an infinite path in $\mathcal{G}(p,r)$ is an infinite pro-p group of coclass r. In particular, there is a one-to-one correspondence between the isomorphism types of these pro-p groups and the coclass trees of $\mathcal{G}(p,r)$. It is Conjecture D of the proven coclass conjectures which says that there are only finitely many isomorphism types of infinite pro-p groups of coclass r, see Section A.2.

We now introduce some more notation to describe coclass trees in more detail. As usual, the depth of a vertex in a rooted tree is its distance from the root and the depth of a rooted tree is the maximum depth of a vertex in the tree. The width of a rooted tree is the maximum number of groups at the same depth.

Definition. Let $\mathcal{T}$ be a coclass tree with corresponding pro-p group S and mainline groups $S_t, S_{t+1}, \ldots$ Let $j \geq t$ and $k \geq 0$ be integers.

a) The j-th branch $\mathcal{B}_j$ of $\mathcal{T}$ is the subtree of $\mathcal{T}$ induced by the descendants of S_j which are not descendants of S_{j+1}.

b) The shaved coclass graph $\mathcal{G}_k(p,r)$ is the subgraph of $\mathcal{G}(p,r)$ induced by the mainline groups of the coclass trees in $\mathcal{G}(p,r)$ and their descendants of distance at most k.

Thus, by construction, the branches of a coclass tree are finite subtrees which are pairwise disjoint, see Figure 1.2. The next step is now to investigate the finitely many coclass trees of the graph $\mathcal{G}(p,r)$. The aim is to draw conclusions from their graph-theoretic structure to the structure of the groups involved. The following conjecture is a weaker version of Conjecture 1.1a) as only the structure of $\mathcal{G}(p,r)$ is considered.

1.2 Conjecture. *Let p be a prime and let r be a positive integer. The coclass graph $\mathcal{G}(p,r)$ can be described by a finite subgraph and periodic patterns.*

1.1. Classification by coclass

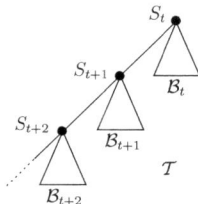

Figure 1.2: The general structure of a coclass tree $\mathcal{T}$.

1.1.3 Shaved coclass graphs and 2-groups of coclass r

Newman & O'Brien considered 2-groups and examined the coclass graph $\mathcal{G}(2,r)$ in some detail. They suggested the conjecture that $\mathcal{G}(2,r)$ can be described by a finite subgraph and periodic patterns, see [42, Conjecture P].

Conjecture P. *If $\mathcal{T}$ is a coclass tree of $\mathcal{G}(2,r)$ with branches $\mathcal{B}_t, \mathcal{B}_{t+1}, \ldots$, then there exists an integer $d = d(\mathcal{T})$ such that, up to finitely many exceptions, for $j \geq t$ the branches $\mathcal{B}_j$ and $\mathcal{B}_{j+d}$ are isomorphic as rooted trees.*

Two years later, in 2001, du Sautoy [9] used zeta functions and model theory to prove a qualitative version of this conjecture. His proof is non-constructive in the sense that one obtains neither information on the periodicity nor on the groups involved.

In 2008, Eick & Leedham-Green [14] published a constructive proof which is based on an explicit group-theoretic construction. They described the groups in a coclass tree $\mathcal{T}$ as suitable group extensions and used cohomology theory to obtain the conjectured graph isomorphisms between the branches of $\mathcal{T}$. Their approach has the advantage that not only one obtains detailed information on the periodicity, but also the branch isomorphisms are reflected in the structure of the groups. It is shown that the 2-groups of coclass r can be described by finitely many parameterized presentations with one integer parameter. This generalizes the well-known result for $r = 1$, see Equation (1.1) on page 2, and proves Conjecture 1.1a) for $p = 2$.

This description by parameterized presentations is of great value as it has several applications. For example, Eick [12] proved that the infinitely many groups which are described by a single parameterized presentation also have automorphism group orders which can be described in a uniform way. Moreover, she conjectured a uniform description of the corresponding Schur multiplicators, see [13]. This supports Conjecture 1.1b) for $p = 2$.

du Sautoy and Eick & Leedham-Green did not only discuss the case $p = 2$ as they also considered shaved coclass graphs in general. They proved that the branches of a coclass tree in a shaved coclass graph are isomorphic with some periodicity.

Theorem P. *Let p be a prime and let r and k be positive integers. Let $\mathcal{T}$ be a coclass tree of the shaved graph $\mathcal{G}_k(p,r)$ with branches $\mathcal{B}_t, \mathcal{B}_{t+1}, \ldots$ Then there exist integers $d = d(\mathcal{T})$ and $f = f(\mathcal{T}, k)$ such that $\mathcal{B}_j$ and $\mathcal{B}_{j+d}$ are isomorphic as rooted trees for all $j \geq f$.*

In a short form, Theorem P says that the shaved coclass graph $\mathcal{G}_k(p,r)$ can be described by a finite subgraph and periodic patterns. Again, Eick & Leedham-Green provided a constructive proof with an explicit value for d and bounds for f.

In general, for odd primes the set of depths of the branches in a coclass tree is unbounded and, by [14, Remark 4], there exists an integer k such that almost all vertices of $\mathcal{G}(p,r)$ are contained in $\mathcal{G}_k(p,r)$ only if either $p = 2$, or both $p = 3$ and $r = 1$. This shows that Conjecture P cannot be adapted to odd primes and the periodic patterns described in Theorem P do not suffice to prove Conjecture 1.2.

Thus, for odd primes, it is necessary to consider the unbounded growth of the branches. Another difficulty is that the number of coclass trees in $\mathcal{G}(p,r)$ grows quickly as r grows. For example, Eick [11, Section 6] showed that $\mathcal{G}(3,3)$, $\mathcal{G}(3,4)$, and $\mathcal{G}(5,3)$ have more than 10^3, 10^{11}, and 10^{16} coclass trees, respectively. On the other hand, it is known that $\mathcal{G}(p,1)$ has exactly one coclass tree for every prime p. Hence, the coclass graph $\mathcal{G}(p,1)$ with $p \geq 5$ is distinguished as a perfect subject of study related to Conjectures 1.1 and 1.2; it seems to be the easiest case for which these conjectures are still open.

1.2 Groups of maximal class

The investigation of maximal class groups started long before coclass theory was developed. Though, despite substantial progress made in the last six decades, the classification of maximal class groups is still an open problem in p-group theory, see Problem 3 of Shalev's survey paper on finite p-groups [51].

Following Blackburn, the theory of maximal class was advanced by Shepherd [52], Miech [35–39], Leedham-Green & McKay [25–28], Fernández-Alcober [17], and Vera-López et al. [24, 57–65]. We already mentioned that the 2- and 3-groups of maximal class are classified. The 5-groups of maximal class were investigated by Leedham-Green & McKay [28] and, computationally, by Newman [41] and Dietrich, Eick & Feichtenschlager [6]. Their results support Conjectures 1.1 and 1.2. There are partial results for primes greater than 5, see [24, 28], but, nonetheless, a classification in the sense of Conjectures 1.1 and 1.2 is still open.

We now follow the philosophy of coclass theory and consider the graph $\mathcal{G}(p) = \mathcal{G}(p,1)$ associated with p-groups of maximal class. It is well-known that $\mathcal{G}(p)$ consists of the cyclic group C_{p^2} of order p^2 and an infinite coclass tree $\mathcal{T}(p)$ with corresponding pro-p group S of coclass 1. The groups on the mainline of $\mathcal{T}(p)$ are $S_2, S_3, \ldots$ where $S_j = S/\gamma_j(S)$ for $j \geq 2$, see Figure 1.3. The coclass trees $\mathcal{T}(2)$ and $\mathcal{T}(3)$ are well-understood. In contrast to these cases, if $p \geq 5$, then the branches of $\mathcal{T}(p)$ have unbounded depth, that is, the set of depths of the branches of $\mathcal{T}(p)$ is unbounded. If $p \geq 7$, then the set of widths of the branches of $\mathcal{T}(p)$ is unbounded, cf. [6, Remark 2.2]. Thus, for $p = 5$, an investigation by computer is still possible, whereas for $p \geq 7$ the size of the branches increases too fast for a complete computational examination.

Coclass theory now proceeds by testing these branches for periodic patterns, and we refer to the results of Newman [41], Dietrich, Eick & Feichtenschlager [6], and Eick, Leedham-Green, Newman & O'Brien [15] for recent computational evidence supporting this approach. The aim of this thesis is to investigate the structure of the coclass tree $\mathcal{T}(p)$ in more detail and to pursue Conjectures 1.1a) and 1.2.

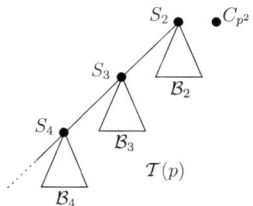

Figure 1.3: The graph $\mathcal{G}(p)$.

1.3 Comment on the notation

Unless otherwise noted, we write all groups multiplicatively. We apply functions from the left and usually write group actions exponentially.

Groups of homomorphisms into abelian groups are written additively; that is, if B is an abelian group (written multiplicatively) and $f, g \colon A \to B$ are homomorphisms, then we apply the homomorphism $f + g$ to an element $a \in A$ via $(f + g)(a) = f(a)g(a)$. In a similar way, we write the group of 1- and 2-cocycles with coefficients in an abelian group as additive groups.

Throughout this thesis, we denote by p a prime and define $d = p - 1$. In most parts, we assume that p is greater than 3. Usually, the symbols n, m, e, k, and r denote non-negative integers. For an overview of the general notation used in this thesis we refer to the list of symbols at the end of this thesis.

2 The graph $\mathcal{G}(p)$

We give a survey of the results achieved in this thesis and, for this purpose, we describe the structure of the coclass graph $\mathcal{G}(p)$ and its coclass tree $\mathcal{T}$ in more detail.

2.1 General notation

As a first step, we settle the notation which is used throughout the thesis and, for the sake of completeness, we recall the relevant definitions made in the introduction.

The vertices of the **coclass graph** $\mathcal{G}(p)$ are the isomorphism types of finite p-groups of maximal class where a vertex is identified with a group representing its isomorphism class. Two vertices G and H are connected by a directed edge $G \to H$ if and only if G is isomorphic to the central quotient $H/\zeta(H)$. A group H in $\mathcal{G}(p)$ is a **descendant** of a group G in $\mathcal{G}(p)$ if G is isomorphic to the quotient $H/\gamma_n(H)$ for some $n \geq 2$; that is, if $G = H$ or if there is a path in $\mathcal{G}(p)$ from G to H. If $n = c(H)$, the nilpotency class of H, then H is an **immediate descendant** of G and G is the **parent** of H. A group having immediate descendants is called **capable** and a non-capable group is **terminal**.

We visualize $\mathcal{G}(p)$ in the Euclidean plane as an undirected graph by drawing the proper descendants of a group in $\mathcal{G}(p)$ below that group.

For a fixed primitive p-th root of unity θ over $\mathbb{Q}_p$ we define

$$P = C_p(\theta) \quad \text{and} \quad T = (\mathbb{Z}_p[\theta], +)$$

where $C_p(\theta)$ is the cyclic group of order p generated by θ, and $\mathbb{Z}_p[\theta]$ is the ring of integers of the p-th local cyclotomic field $\mathbb{Q}_p(\theta)$. Closely related to $\mathcal{G}(p)$ is the **p-adic space group**

$$S = P \ltimes T$$

where P acts via multiplication by θ on T. Up to isomorphism, the group S is the unique infinite pro-p group of coclass 1. The groups P and T are the **point group** and **translation subgroup** of S, and for $n \geq 2$ we write $T_1 = T$ and

$$T_n = \gamma_n(S) \quad \text{and} \quad S_n = S/\gamma_n(S).$$

It is well-known and proved in Theorem 6.4 that the graph $\mathcal{G}(p)$ consists of the cyclic group of order p^2 and an infinite **coclass tree** $\mathcal{T} = \mathcal{T}(p)$ with elementary abelian root of order p^2. The groups on the **mainline** of $\mathcal{T}$ are the **mainline groups** $S_2, S_3, \ldots$

For $n \geq 2$, the finite subtree $\mathcal{B}_n$ of $\mathcal{T}$ induced by all descendants of S_n which are not descendants of S_{n+1} is the n-**th branch** of $\mathcal{T}$. For a group G in $\mathcal{T}$ and a positive integer k the **k-step descendant tree** $\mathcal{D}_k(G)$ of G is the subtree of $\mathcal{T}$ induced by the descendants of distance at most k from G. If $\mathcal{B}$ is any subtree of $\mathcal{T}$ with root G, then the **shaved subtree** $\mathcal{B}[k]$ is the subtree of $\mathcal{B}$ induced by the groups of distance at most k from G.

The 2- and 3-groups of maximal class are classified and, thus, we assume that p is a prime greater than 3 throughout the thesis.

2.1.1 Twigs, body, and skeleton

In this paragraph we introduce some important subtrees of a branch; namely, the body, the skeleton, and the twigs of a branch. First, we need some more notation.

2.1 Definition. Depending on p, we define the following integer constants.
a) Let $d = p - 1$ be the dimension of S, that is, the $\mathbb{Z}_p$-rank of its translation subgroup.
b) Let $\mathfrak{c} = \mathfrak{c}(p)$ be $4p - 19$ if $p \geq 7$, and 4 if $p = 5$.
c) For $n \geq 2$ let $\mathfrak{e}_n = \mathfrak{e}_n(p)$ be $\max\{0, n - 2p + 8\}$ if $p \geq 7$, and $\max\{0, n - 4\}$ if $p = 5$.

The following definition has been motivated by Leedham-Green & McKay, see [29, Definition 8.4.8].

2.2 Definition. Let $f\colon T \wedge T \to T_n$ be a surjective P-homomorphism and let $e \leq \lfloor \frac{n-1}{d} \rfloor d$ be a positive integer. Then f defines a new multiplication on the additive group T/T_{n+e} via
$$(a + T_{n+e}) \odot (b + T_{n+e}) = a + b + \tfrac{1}{2}f(a \wedge b) + T_{n+e} \quad (a, b \in T),$$
and the constructible group or skeleton group defined by f and e is
$$C_{f,e} = P \ltimes (T/T_{n+e}, \odot)$$
where P acts via multiplication by θ. A group in the branch $\mathcal{B}_n$ is a skeleton group if it is isomorphic to $C_{h,e}$ for some e and some surjective P-homomorphism $h\colon T \wedge T \to T_n$.

We consider skeleton groups in more detail in Section 7.3 and show that the next definition is well-defined.

2.3 Definition. Let $n \geq 2$ be an integer.
a) The body $\mathcal{T}_n$ of $\mathcal{B}_n$ is the shaved branch $\mathcal{B}_n[\mathfrak{e}_n]$.
b) The skeleton $\mathcal{S}_n$ of $\mathcal{B}_n$ is the subtree of $\mathcal{T}_n$ induced by the skeleton groups in $\mathcal{T}_n$.
c) The groups in a body not contained in the skeleton are called twig groups. Together with their parents in $\mathcal{T}_n$, they form trees with roots in the skeleton of $\mathcal{T}_n$ called twigs.

We call $\mathcal{T}_n$ and $\mathcal{S}_n$ the n-**th body** and n-**th skeleton**, respectively, of the tree $\mathcal{T}$.

The definition of the skeleton is motivated by the explicit construction of the groups in Definition 2.2. We prove later in Lemma 6.6 and Section 7.1 that the groups in the n-th body $\mathcal{T}_n$ can be described in a practical way as group extensions. This is the basis for our further investigations and thus motivates the definition of a body.

2.2 Bounding the depths

As usual, the **depth of a vertex** in a rooted tree is its distance from the root, and the **depth of a rooted tree** is the maximum depth of a vertex in the tree. The **width of a rooted tree** is the maximum number of groups at the same depth. We now investigate the depths of a branch, its body, skeleton, and twigs. The following theorem is proved in Corollaries 7.20 and 7.23 and Theorem 7.21.

2.4 Theorem. *Let $n \geq p+1$ be an integer.*
a) *The depth of the branch $\mathcal{B}_n$ is at most $\mathfrak{e}_n + \mathfrak{c}$.*
b) *The depth of the body $\mathcal{T}_n$ and the depth of the skeleton $\mathcal{S}_n$ both are $\mathfrak{e}_n$.*
c) *The twigs in the body $\mathcal{T}_n$ are trees of depth 1.*

The bound on the depths of the branches of $\mathcal{T}(5)$ is sharp, see Section 10.1. For $p = 7$ computer experiments suggest that the branch $\mathcal{B}_n$ has depth at most $n+2$; that is, in general, the bound in Theorem 2.4a) does not seem not to be sharp.

Theorem 2.4 has some important consequences. First, it shows that the body of a branch is a significant subtree: On the one hand, the difference of the depths of a branch $\mathcal{B}_n$ and its body $\mathcal{T}_n$ is bounded by the constant $\mathfrak{c}$ for all $n \geq p+1$. On the other hand, the ratio of these depths converges to 1 as n tends to infinity. Second, it follows from part c) that the general shape of a body is determined completely by its skeleton. The construction of skeleton groups is given explicitly in Definition 2.2 and a strategy to construct these groups up to isomorphism is in principle developed by Leedham-Green & McKay [28]. But, as they say, in practice it is a complicated procedure which relies on subtle number theory. We elaborate the underlying number theory and report on the corresponding problems in Section 5.4 and Chapter 9.

The next corollary can be deduced from results of Leedham-Green & McKay [28, 29], cf. [6, Remark 2.2]. Here we prove it as a consequence of Theorem 2.4, Corollary 9.4, and Theorem 10.1. Recall that the branches of $\mathcal{T}(p)$ have (un)bounded width or depth if the set of widths or depths of the branches is (un)bounded, respectively.

2.5 Corollary. a) *The branches of $\mathcal{T}(2)$ and $\mathcal{T}(3)$ have bounded width and depth.*
b) *The branches of $\mathcal{T}(5)$ have bounded width and unbounded depth.*
c) *The branches of $\mathcal{T}(p)$ with $p \geq 7$ have unbounded width and depth.*

2.3 Periodicity of type 1

The first main result of this thesis is a strengthened version of Theorem P, page 5, for the coclass tree $\mathcal{T}$ of $\mathcal{G}(p)$; the following theorem is proved in Theorem 8.4.

2.6 Theorem. *If $n \geq p+1$, then there is an embedding*
$$\iota = \iota_n \colon \mathcal{T}_n \hookrightarrow \mathcal{B}_{n+d}$$
of rooted trees such that $\iota(\mathcal{T}_n) = \mathcal{B}_{n+d}[\mathfrak{e}_n]$.

We refer to the periodic pattern described in Theorem 2.6 as the periodicity of type 1; it is sketched in Figure 2.1. This periodicity has been conjectured, due to computer experiments, for a long time, but it has never been proved before. It is motivated by the results of du Sautoy [9] and Eick & Leedham-Green [14], as well as by detailed computer experiments for $p = 5$, see Newman [41] and Dietrich, Eick & Feichtenschlager [6].

In the case of coclass 1, Theorem P asserts that for every positive integer k there is an integer $f = f(k)$ such that the shaved branches $\mathcal{B}_n[k]$ and $\mathcal{B}_{n+d}[k]$ are isomorphic as rooted trees whenever $n \geq f$. Eick & Leedham-Green [14] determined an upper bound for the least possible value of f and proved that
$$f(k) \leq (6d + 3/2)k + (6d + 15/2)d + 1.$$

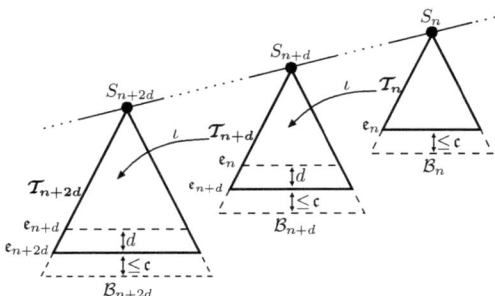

Figure 2.1: The periodicity of type 1.

Using this bound, it follows from Theorem P that a shaved subtree of the body $\mathcal{T}_n$ of depth approximately $n/(6d)$ can be embedded into $\mathcal{B}_{n+d}$. In contrast, the periodicity mapping ι_n of Theorem 2.6 embeds the whole body $\mathcal{T}_n$ of depth $\mathfrak{e}_n \approx n$ into $\mathcal{B}_{n+d}$. Together with Theorem 2.4, this shows that the periodicity of type 1 proved here is a stronger version of the periodicity of $\mathcal{T}$ proved by Eick & Leedham-Green [14] and du Sautoy [9].

Computational investigations reveal that it is not possible to embed the whole branch $\mathcal{B}_n$ into $\mathcal{B}_{n+d}$. However, it follows from Theorems 2.4 and 2.6 that $\mathcal{B}_n$ can be embedded with the exclusion of at most $\mathfrak{c}$ levels of groups whenever $n \geq p+1$. This indicates that the periodicity mappings of Theorem 2.6 are close to best possible. Though, experiments suggest that the largest subtree of $\mathcal{B}_n$ which can be embedded into $\mathcal{B}_{n+d}$ might be bigger than the body $\mathcal{T}_n$. For $p = 5$ it is conjectured that $\mathcal{B}_n[n-1] \cong \mathcal{B}_{n+4}[n-1]$ for all $n \geq 6$.

We remark that the proof of Theorem 2.6 given here yields an alternative and independent proof for the periodicity of $\mathcal{T}$ considered in [9] and [14].

Definition. Let $n \geq p+1$. The periodicity class $\mathcal{P}(G)$ of a group G in the body $\mathcal{T}_n$ is defined as the infinite sequence of groups

$$\mathcal{P}(G) = (G, \iota(G), \iota^2(G), \ldots).$$

As in [14], we prove that the graph isomorphisms induced by Theorem 2.6 are reflected in the structure of the groups; that is, the embeddings can be chosen such that the groups in a periodicity class can be described in a uniform way.

2.7 Theorem. *Let $n \geq p+1$. If G is a group in the body $\mathcal{T}_n$, then the groups in the periodicity class $\mathcal{P}(G)$ can be described by a single parameterized presentation with one integer parameter.*

Theorems 2.6 and 2.7 strongly support Conjecture 1.1a) for maximal class groups. An example of explicit parameterized presentations and a proof of Theorem 2.7 are given in Theorem 8.6.

2.4 Periodicity of type 2

We have shown that the body $\mathcal{T}_n$ with $n \geq p+1$ can be embedded into $\mathcal{B}_{n+d}$ such that $\mathcal{T}_n \cong \mathcal{B}_{n+d}[\mathfrak{e}_n]$ as rooted trees. In order to describe the branch $\mathcal{B}_{n+d}$ completely, it is necessary to describe its growth, that is, the difference graph of $\mathcal{B}_{n+d}$ and $\iota_n(\mathcal{T}_n) = \mathcal{B}_{n+d}[\mathfrak{e}_n]$. Since in general the widths of the branches are unbounded, this graph cannot be isomorphic to a subgraph of $\mathcal{B}_n$. However, a conjecture of Eick, Leedham-Green, Newman & O'Brien [15] claims that it can be described by another periodic pattern, that is, a periodicity of type 2. According to this conjecture, for large enough n, the descendant tree of a group at depth $\mathfrak{e}_n$ in $\mathcal{B}_{n+d}$ is isomorphic to a descendant tree of a group at depth $\mathfrak{e}_{n-d}$ in $\mathcal{B}_n$.

As a first approximation of this conjecture, we consider the difference graph of $\mathcal{T}_{n+d}$ and $\iota(\mathcal{T}_n)$ in this thesis; that is, we restrict attention to the bodies of $\mathcal{T}$ and, thus, ignore at most $\mathfrak{c}$ levels of groups per branch. A periodicity of type 2 for the bodies of $\mathcal{T}$ is conjectured as follows, see Figure 2.2. Recall that $\mathcal{D}_d(G)$ is the d-step descendant tree of a group G in $\mathcal{T}$.

2.8 Conjecture. *There is an integer $n_0 = n_0(p)$ with the following property: If $n \geq n_0$ and G is a group at depth $\mathfrak{e}_n$ in $\mathcal{T}_{n+d}$, then there exists a group H at depth $\mathfrak{e}_{n-d}$ in $\mathcal{T}_{n+d}$ such that*

$$\mathcal{D}_d(G) \cong \mathcal{D}_d(H)$$

as rooted trees. The group H is called a periodic parent of G.

The main problem is to specify a periodic parent of a given group, and it is expected that a proof of Conjecture 2.8 yields a *natural* mapping which chooses such periodic parents. Hence, the periodicities of type 1 and 2 would suffice to describe the bodies of $\mathcal{T}$ completely by a finite subgraph and periodic patterns, which supports Conjecture 1.2, cf. [15].

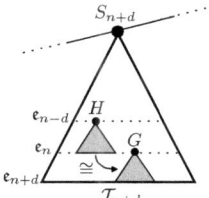

Figure 2.2: A periodic parent of G.

It follows from Corollary 9.2 that the degrees of the vertices in the tree $\mathcal{T}$ are bounded by a constant and, thus, there are only finitely many trees which can occur as a d-step descendant tree of a group in $\mathcal{T}$. Therefore, from a graph theoretic point of view, it seems reasonable to propose Conjecture 2.8. Though, the group G and a periodic parent $\Pi(G)$ should not only have isomorphic d-step descendant trees, which holds by definition, but also the periodic parent should be chosen carefully such that the groups in the d-step descendant tree of G can be constructed somehow from the groups in the d-step descendant tree of $\Pi(G)$; such a construction would eventually allow the definition of additional periodicity classes which would support Conjecture 1.1a).

In this thesis, we investigate the periodicity of type 2 with respect to Conjecture 1.2. The following theorem is proved in Theorem 9.10, cf. Figure 2.3. Recall that the p'-part of an integer u is the integer v with $p \nmid v$ and $u = vp^a$ for some integer $a \geq 0$.

2.9 Theorem. *Let $p \equiv 5 \bmod 6$ and let G be a group at depth $\mathfrak{e}_n$ in the body $\mathcal{T}_{n+d}$. There exists a positive integer $n_0 = n_0(p)$ such that for all $n \geq n_0$ the following holds. If there is a maximal path $K_0 \to K_1 \to \ldots$ in $\mathcal{T}_{n+d}$ such that G has distance $k \leq d$ from a group on this path and if the p'-part of the order of $\mathrm{Aut}(K_s)$ is the same for all $\mathfrak{e}_{n-d} - k \leq s \leq \mathfrak{e}_{n+d}$, then G has a periodic parent $\Pi(G)$ of distance k from this path.*

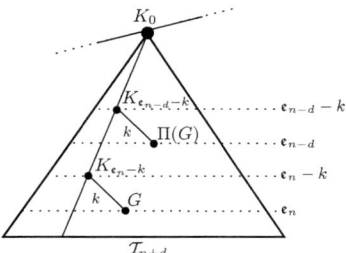

Figure 2.3: The group G and a periodic parent $\Pi(G)$.

Our proof of Theorem 2.9 is constructive in the sense that we give an explicit isomorphism between the d-step descendant trees of G and $\Pi(G)$ on a group theoretic level; that is, we describe how the groups in $\mathcal{D}_d(G)$ arise from those in $\mathcal{D}_d(\Pi(G))$. Note that Conjecture 2.8 is proved for $p \equiv 5 \bmod 6$ if the assumptions of Theorem 2.9 are always satisfied. So far, we are not able to contribute to a positive solution of this problem.

In Corollary 9.11, we provide a bound for the integer n_0 in terms of p and μ, an integer which depends on some p-adic valuations of certain p-adic integers. Computer experiments with these p-adic valuations suggest that $n_0 \leq d(d/2 + 8) - 6$ if $p \geq 7$ and $n_0 \leq 32$ if $p = 5$. These bounds seem not be sharp, cf. Section 10.1.

The following corollary is a consequence of Theorem 2.9; it is proved in Corollary 9.12. Note that the d-**step parent** of a group G in $\mathcal{G}(p)$ of order p^n with $n > p$ is $G/\gamma_{n-d}(G)$.

2.10 Corollary. *Let $p \equiv 5 \bmod 6$ and $n \geq n_0$. Let G be a capable group at depth $\mathfrak{e}_n$ in the body $\mathcal{T}_{n+d}$ and let H be the d-step parent of G. If the automorphism group of H is a p-group, then H is a periodic parent of G.*

Corollary 2.10 seems to cover an important special case as the majority of the groups in $\mathcal{G}(p)$ seem to have a p-group as automorphism group. For example, it is conjectured by Newman [41] that 90% of all groups in $\mathcal{G}(5)$ have a 5-group as automorphism group, cf. [34]. This indicates that a significant part of the growth of $\mathcal{T}_{n+d}$ can be described by using d-step parents as periodic parents. However, computer experiments suggest that there exist infinitely many n such that there are capable groups at depth $\mathfrak{e}_n$ in $\mathcal{T}_{n+d}$ whose d-step parent is *not* a periodic parent. A remark supporting this observation is given in Section 9.3.3.

Our proof of Theorem 2.9 is based on ideas of Leedham-Green & McKay [28] to construct skeleton groups up to isomorphism and thus uses p-adic number theory. As suggested by Leedham-Green & McKay, there are significant differences depending on the residue class p modulo 6, and there occur additional problems if $p \equiv 1 \bmod 6$. Hence, we are not able to adapt the previous results to primes $p \equiv 1 \bmod 6$. However, computer experiments for $p = 7$ suggest that the situation in general is similar to $p = 5$; that is, for the majority of capable groups the d-step parent is a periodic parent, but there also seem to be infinitely many cases where this does not hold.

2.5 The graph $\mathcal{G}(5)$

Using the previous results and the investigation of so-called one-parameter groups by Leedham-Green & McKay [28], we prove the following theorems for the coclass tree $\mathcal{T}$ of $\mathcal{G}(5)$, see Section 10.1.

Theorem. *If G is a capable group at depth $\mathfrak{e}_n$ in $\mathcal{T}_{n+4}$ with $n \geq 14$, then the 4-step parent of G is a periodic parent.*

Theorem. *The groups in the bodies of $\mathcal{T}$ can be described by finitely many parameterized presentations with at most two integer parameters.*

These results strongly support Conjectures 1.1a) and 1.2, and they are close to a positive answer of Problem 3 of Shalev's survey paper [51]. Moreover, the computational investigation of $\mathcal{G}(5)$ by Dietrich, Eick & Feichtenschlager [6] supports Conjecture 1.1b). However, the coclass tree $\mathcal{T}$ has bounded width and this distinguishes $\mathcal{T}$ from the coclass tree $\mathcal{T}(p)$ with $p > 5$. The structure of 5-groups of maximal class is much more restricted than the structure of p-groups of maximal class for $p > 5$.

2.6 Open problems

There are three problems to be solved in order to prove Conjecture 1.2 for the graph $\mathcal{G}(p)$.

1. Complete the proof of Conjecture 2.8 for $p \equiv 5 \bmod 6$: Either, it is to show that the assumptions of Theorem 2.9 are always satisfied for large enough n, or one has to construct periodic parents for groups which are not covered by Theorem 2.9. As indicated in Theorem 2.9, the p'-parts of the orders of some automorphism groups play an important role.

2. Prove Conjecture 2.8 for $p \equiv 1 \bmod 6$: If $p \equiv 1 \bmod 6$, then there are additional difficulties in the number theory used to construct skeleton groups up to isomorphism. Hence, an alternative proof is needed.

3. Consider the groups in the difference graph of the branch $\mathcal{B}_n$ and its body $\mathcal{T}_n$: Throughout the thesis, we restrict attention to groups in the body $\mathcal{T}_n$ as they can be described in a practical way as certain group extensions. This is not possible for all groups in the difference graph of $\mathcal{B}_n$ and $\mathcal{T}_n$.

However, the question whether Conjectures 1.1 and 1.2 hold for coclass 1 is still open and, thus, it is not obvious that the above problems can be solved. We think it is difficult to assess the situation. On the one hand, for $p \geq 7$ it is hard to compute significant examples as the branches of the coclass tree $\mathcal{T}(p)$ are too complex for a complete computational investigation. For example, the branches $\mathcal{B}_7$ and $\mathcal{B}_8$ of the tree $\mathcal{T}(7)$ already contain 20295 and 317830 capable groups, respectively. Although naive random computer experiments suggest that the d-step parent is a periodic parent for the majority of groups, the observation that there seem to be infinitely many exceptions is a setback. In general, it remains to describe another construction of periodic parents, and Theorem 2.9 is already a step ahead in solving this challenge. On the other hand, there *is* strong computational evidence that the graph $\mathcal{G}(p)$ can be described by periodic patterns and, moreover, the new periodicities proved in this thesis indicate that there are a lot of recurrent patterns in the graph $\mathcal{G}(p)$.

2.6.1 Generalizations

With respect to Conjectures 1.1 and 1.2, it might be of interest to set the results of this thesis in the context of arbitrary coclass. In particular, a periodicity of type 2 is conjectured by Eick, Leedham-Green, Newman & O'Brien [15] for coclass trees of arbitrary coclass graphs and, therefore, one could ask for a generalization of Theorem 2.9.

There are two aspects of an answer to this question. On the one hand, the proofs in this thesis substantially make use of the specific properties of maximal class groups; that is, they are based explicitly on theory and methods connected to maximal class. This, of course, shows that one cannot expect a generalization by simply transferring the proofs. On the other hand, the cohomological approach used in this thesis is motivated by Eick & Leedham-Green [14] who successfully used cohomology to investigate (shaved) coclass trees for arbitrary coclass. Thus, cohomology seems to be a fruitful approach and we think that one could in principle extend the methods used in this thesis to arbitrary coclass. This, of course, would require much more effort – in particular, because the case of maximal class has not been solved completely yet.

3 Basics of maximal class groups

In this chapter we recall some basic properties of maximal class groups. We introduce the degree of commutativity and point out the connection between maximal class and uniserial actions. Most results of this chapter are well-known and can be found in [29].

3.1 Normal subgroup structure

First, we fix the notation of commutators as we use it throughout the thesis.

Definition. Let G be a group and $z_1, \ldots, z_n \in G$ with $n \geq 3$.
a) We define $[z_1, z_2] = z_1^{-1} z_2^{-1} z_1 z_2$ and, by induction, $[z_1, \ldots, z_n] = [[z_1, \ldots, z_{n-1}], z_n]$. If $z_2 = \ldots = z_n = z$, then $[z_1, \ldots, z_n]$ is abbreviated by $[z_1,_{n-1} z]$.
b) If $X, Y \subseteq G$, then $[X, Y]$ is the subgroup of G generated by $\{[x, y] \mid x \in X, y \in Y\}$.

Recall that a maximal class group G of order p^n has lower central series
$$G = \gamma_1(G) > \ldots > \gamma_n(G) = \{1\}$$
where $\gamma_{j+1}(G) = [\gamma_j(G), G]$ for $j \geq 1$. All factors have order p, except the first one which has order p^2 and which is elementary abelian if $n \geq 3$. The following lemma shows that the terms of the lower central series of a maximal class group determine the normal subgroup structure, see [29, Proposition 3.1.2].

3.1 Lemma. *If G has maximal class and $N \trianglelefteq G$ has index p^r with $r \geq 2$, then $N = \gamma_r(G)$. The group G has $p + 1$ normal subgroups of index p.*

Proof. The quotient G/N has order p^r and nilpotency class less than r, that is, $\gamma_r(G) \leq N$, and $N = \gamma_r(G)$ follows from $[G : \gamma_r(G)] = p^r$. The group G is nilpotent and hence the normal subgroups of index p are exactly the maximal subgroups of G. The factor $G/\gamma_2(G)$ is elementary abelian and thus $\gamma_2(G) = \Phi(G)$, the Frattini subgroup of G. Therefore, G has $p + 1$ maximal subgroups. □

The maximal class groups of order p^n with $n = 2, 3$ can be determined readily. There are two (isomorphism types of) groups of order p^2 and both have maximal class. Also, there are two maximal class groups of order p^3; they can be described by the group presentations

$$\langle a, b, c \mid a^p = 1, b^p = 1, c^p = 1, b^a = bc, c^a = c, c^b = c \rangle \quad \text{and}$$
$$\langle a, b, c \mid a^p = c, b^p = c, c^p = 1, b^a = bc, c^a = c, c^b = c \rangle.$$

If n is greater than 3, then the lower central series can be refined to a composition and chief series by adding an additional term.

Definition. The refined (lower) central series of a maximal class group G of order p^n with $n \geq 4$ is
$$G > P_1(G) > \ldots > P_n(G) = \{1\}$$
where $P_j(G) = \gamma_j(G)$ if $j \geq 2$, and $P_1(G) = C_G(P_2(G)/P_4(G))$ is the 2-step centralizer of G. If there is no risk of confusion, then we write P_j for $P_j(G)$.

If we consider a maximal class group with refined central series $G > P_1 > \ldots > P_n = \{1\}$, then we implicitly assume that $n \geq 4$.

3.2 Degree of commutativity

An important invariant of a maximal class group is its degree of commutativity. It measures how far down commutation goes inside the 2-step centralizer.

Definition. If G has maximal class and refined central series $G > P_1 > \ldots > P_n = \{1\}$, then the degree of commutativity $\mathrm{doc}(G)$ of G is the maximum integer l such that $[P_i, P_j] \leq P_{i+j+l}$ for all $i, j \geq 1$ if P_1 is not abelian, and $\mathrm{doc}(G) = n - 3$ if P_1 is abelian.

We now provide a lower bound for the degree of commutativity. For a proof, which relies on technical commutator calculations, we refer to [17] and [29, Theorem 3.3.5 & p. 69].

3.2 Theorem. *Let G be a maximal class group of order p^n with $n \geq 4$.*
a) *If $n > p + 1$, then $\mathrm{doc}(G)$ is positive.*
b) *If $p \geq 7$, then $\mathrm{doc}(G)$ is at least $(n - 2p + 5)/2$.*
c) *If $p = 2, 3, 5$, then $\mathrm{doc}(G)$ is at least $n - 2$, $n - 4$, and $(n - 6)/2$, respectively.*
d) *If $n \geq 5$, then $\mathrm{doc}(G/P_n(G)) < \mathrm{doc}(G)$ if and only if $P_1(G)$ is abelian.*

The following lemma summarizes some results of Lemmas 3.2.4, 3.3.1, and 3.3.7 in [29]. For a group G and a subset $M \subseteq G$ we denote by $\langle M \rangle$ the subgroup of G generated by M. If $M = \{g_1, \ldots, g_n\}$ is finite, then we also write $\langle M \rangle = \langle g_1, \ldots, g_n \rangle$.

3.3 Lemma. *Let G be of maximal class with refined central series $G > P_1 > \ldots > P_n = \{1\}$. Let $s \in G \setminus P_1$ and $s_1 \in P_1 \setminus P_2$, and $s_j = [s_{j-1}, s] = [s_{1,j-1} s]$ for $j \geq 2$. If G has positive degree of commutativity, then the following holds.*
a) *The group G is generated by $\{s, s_1\}$, and $P_j = \langle s_j \rangle P_{j+1}$ for all $1 \leq j \leq n - 1$.*
b) *If $t \in G \setminus P_1$, then $t^p \in P_{n-1}$.*
c) *If $t \in P_2$, then st is a conjugate of s, and so $(st)^p = s^p$.*

Proof. a) The first assertion follows from $P_2 = \Phi(G)$. For $j \geq 2$ it is a standard result that $P_j = \langle \{[x_1, \ldots, x_j] \mid x_i \in \{s, s_1\}\} \rangle P_{j+1}$, see [29, Lemma 1.1.23]. If $1 < i < n - 1$, then $[s_{1,i-1} s, s_1] \in P_{i+2}$ as G has positive degree of commutativity. This implies the assertion.
b) The quotient G/P_2 has exponent p and thus t^p lies in P_2. Assume, for a contradiction, that $t^p \in P_r \setminus P_{r+1}$ for some $r \leq n - 2$. Then $t^p P_{r+1}$ generates P_r/P_{r+1} and, as t acts trivially on P_{r+1}/P_{r+2}, it follows that $t \in K$ with $K = C_G(P_r/P_{r+2})$. The positive degree of commutativity implies that $P_1 \leq K$ and part a) shows that $K \neq G$. The group P_1 has index p in G and hence $P_1 = K$. Now $t \in P_1$ is a contradiction.

c) It follows from $G = \langle s \rangle P_1$ that every conjugate of s in G is a conjugate of s by an element of P_1. The centralizer of s in P_1 is the center $\zeta(G) = P_{n-1}$, and so s has p^{n-2} conjugates in P_1. Note that $[s,g] \in P_2$ for all $g \in G$ and so $s^g \in sP_2$. Now $|sP_2| = p^{n-2}$ shows that every element of sP_2 is a conjugate of s in G as required. Clearly, $(s^g)^p = (s^p)^g = s^p$ as s^p is central. $\square$

3.3 Maximal class and uniserial action

If G is a group of maximal class with refined central series $G > P_1 > \ldots > P_n = \{1\}$, then
$$P_{i+1} = [P_i, G] \quad \text{and} \quad [P_i : P_{i+1}] = p$$
for all $1 \leq i \leq n-1$; note that $P_2 = [G,G] = [P_1, G]$. We say that G **acts uniserially** on P_1 and we now generalize this concept.

Let G and N be groups. The group N together with a homomorphism $\varphi \colon G \to \mathrm{Aut}(N)$ is a **G-group** where $g \in G$ acts on $n \in N$ via $n^g = \varphi(g)(n)$. An abelian G-group N is called a **G-module**. For $g \in G$ and $n \in N$ a commutator is defined as
$$[n, g] = n^{-1} n^g \in N.$$
Analogously, $[H, G] = \langle \{[h,g] \mid h \in H, g \in G\} \rangle$ for any G-invariant subgroup $H \leq N$.

Definition. Let N and G be finite p-groups with G acting on N.
a) For a G-invariant subgroup $H \leq N$ let $H_1 = H$ and, inductively, $H_i = [H_{i-1}, G]$.
b) The group G acts uniserially on N if $[H_1 : H_2] = p$ for every nontrivial G-invariant subgroup $H \leq N$.

If G acts uniserially on N and $H \leq N$ is a G-invariant subgroup, then there exists an integer m with $H_m = \{1\}$ such that $H = H_1 > \ldots > H_m = \{1\}$ is a subnormal series of G-invariant subgroups. The following lemma shows that *all* G-invariant subgroups of N can be linearly ordered. The proof is from [29, Lemma 4.1.3 & Corollary 4.1.4].

Lemma. *Let G be a finite p-group and let N be a G-group of order p^n. Then G acts uniserially on N if and only if $N_n \neq \{1\}$; and so $N = N_1 > \ldots > N_{n+1} = \{1\}$, and $[N_i : N_{i+1}] = p$ for $1 \leq i \leq n$. If G acts uniserially on N, then the subgroups N_i for $1 \leq i \leq n+1$ are all the G-invariant subgroups of N and they are normal in N.*

Proof. First, we prove that if G acts on N with series $N = N_1 > N_2 > \ldots > N_{n+1} = \{1\}$, then the subgroups $N_1, \ldots, N_{n+1}$ are all the G-invariant subgroups of N. For $n = 1$ the result is trivial, and so assume that $n \geq 2$ and that the result is true for $n-1$. Let $M < N$ be a G-invariant subgroup of N; that is, $M \cap N_2 = N_i$ for some $i \geq 2$ by the inductive hypothesis applied to N_2. If $M \leq N_2$, then $M = N_i$. If M is not contained in N_2, then $MN_2 = N$ and $[M, G] \leq M \cap N_2 = N_i$ with $i > 2$; that is, $N_2 = [MN_2, G] = [M, G][N_2, G] = N_3$, giving a contradiction.

Now, if $N_n \neq \{1\}$, then $N = N_1 > \ldots > N_{n+1} = \{1\}$, and these are all G-invariant subgroups of N. This shows that G acts uniserially. Conversely, if G acts uniserially, then $[N_i : N_{i+1}] = p$ for all $1 \leq i \leq n$, and so $N_n \neq \{1\}$. We use induction on n to show that $N_1, \ldots, N_{n+1}$ are normal in N. The center $\zeta(N)$ of N is G-invariant and, thus, $N_n \leq \zeta(N)$; that is, N_n is normal. Now the inductive hypothesis can be applied to N/N_n. $\square$

If N is a G-group, then the **split extension** $G \ltimes N$ is the group with underlying set $G \times N$ and multiplication $(u,n)(v,m) = (uv, n^v m)$; if there is no risk of confusion, then we freely identify $G \times \{1\} = G$ and $\{1\} \times N = N$ and write un for the element (u,n).

3.4 Lemma. *Let $G = P \ltimes N$ be a finite p-group where $P \cong C_p$. Then G has maximal class if and only if G acts uniserially on N.*

Proof. If $N_1 = N$, then $N_2 = [N_1, G] = [G, G] = \gamma_2(G)$. Now induction proves that $N_i = \gamma_i(G)$ for all $i \geq 2$. □

We end this section with two preliminary lemmas, which are used later in the construction of maximal class groups. The second lemma is a modification of [29, Lemma 4.1.16].

A group M is a **central extension** of an abelian group A by a group B if there is a normal subgroup $N \leq \zeta(M)$ with $M/N \cong B$ and $N \cong A$; we usually identify A with N.

3.5 Lemma. *Let M be a finite central extension of a p-group A by an abelian p-group B. We consider $P \cong C_p$ acting on M such that A is P-invariant, and we define $E = P \ltimes M$.*
a) *If $A \leq N \leq M$ and P acts uniserially on N, then E acts uniserially on N.*
b) *We assume that P acts uniserially on A and M/A with $|M/A| \geq p^3$. Let N/A be the minimal P-invariant subgroup of M/A. If E has maximal class and positive degree of commutativity, then P acts uniserially on N.*

Proof. Note that $[M, M] \leq A$ and $[m, qn] = [m, n][m, q]^n$ for all $m, n \in M$ and $q \in P$.
a) Let $N = N_1 > \ldots > N_m = \{1\}$ be the P-uniserial series of N with $A = N_k$ for some k; that is, $N_{i+1} = [N_i, P] = [N_i, E]$ for all $i \geq k$. If $i < k$, then $A \leq [N_i, P] \leq [N_i, E]$ and $N_{i+1} = [N_i, E]$ follows from $N_{i+1}/A = [N_i, P]/A = [N_i, E]/A$.
b) Let $E > P_1 > \ldots > P_n = \{1\}$ be the refined central series of E. It follows from Lemma 3.1 that $A = P_{k+1}$ for some $k \geq 3$, and $P_1 = M$ since $P_2 \leq M$ and $A \leq P_4$. By Lemma 3.4, the group E acts uniserially on P_1 and, hence, $N = P_k$. The P-uniserial series of A is $A = P_{k+1} > \ldots > P_n = \{1\}$, and it remains to show that $A = [N, P]$. This follows from $A = [P_k, E] = [P_k, M][P_k, P]$ and $[P_k, M] \leq [P_k, P]$ since $M = P_1$ and E has positive degree of commutativity. □

3.6 Lemma. *Let E and N be finite p-groups with E acting on N. Let $N \geq C > A \geq D$ be E-invariant normal subgroups of N such that E acts uniserially on N/A and C/D. If every E-invariant subgroup $A \geq U \geq D$ is normal in N, then E acts uniserially on N/D.*

Proof. We use induction on $[A : D]$ and consider $A > D$. If $[A : D] > p$, then there is an E-invariant $A > U > D$ and the inductive hypothesis applied to $N \geq C > A > U$ shows that E acts uniserially on N/U. Now we can apply the inductive hypothesis to $N \geq C > U \geq D$ and obtain that E acts uniserially on N/D, as required. It remains to consider $[A : D] = p$. Assume, for a contradiction, that U/D is a minimal E-invariant subgroup of N/D not equal to A/D; that is, $U \cap A = D$. Then $N \geq UA \geq A$ and $UA/A \cong U/D$ as E-groups. Hence, UA/A is a minimal E-invariant subgroup of N/A and $C \geq UA$; that is, $C \geq U$ and $C/D \geq U/D$. This shows that $U = A$ as A/D is the unique minimal E-invariant subgroup of C/D, a contradiction. Thus A/D is the unique minimal E-invariant subgroup of N/D. As E acts uniserially on N/A and C/D, this proves the lemma. □

4 Polycyclic groups and cohomology

Every maximal class group has a subnormal series with cyclic factors. A group with this property is called polycyclic and can be described efficiently by a group presentation on a set of generators chosen through this series. In the first part of this chapter, we recall some relevant properties of polycyclic groups and presentations; we refer to [10, 22, 54] for proofs and further information. The second half is devoted to group cohomology and we recall the connection between group extensions and second cohomology groups, see [23, 47]. In particular, we describe how to compute second cohomology groups of polycyclic groups. These results are motivated by and based on [4, 16, 22].

4.1 Polycyclic groups

A group G is **polycyclic** if it admits a **polycyclic series**; that is, a subnormal series

$$G = G_1 \geq \ldots \geq G_{n+1} = \{1\}$$

such that every factor G_i/G_{i+1} is cyclic. It is easy to see that G is polycyclic if and only if G has an ordered generating set $(h_1, \ldots, h_m)$ such that the groups

$$H_i = \langle h_i, \ldots, h_m \rangle, \quad 1 \leq i \leq m,$$

with $H_{m+1} = \{1\}$ form a polycyclic series of G. In this case, $(h_1, \ldots, h_m)$ is called a **polycyclic sequence** of G. The **relative orders** of its elements are $(r_1, \ldots, r_m)$ where $r_i = \mathrm{relord}(h_i)$ is the order of the factor H_i/H_{i+1}. The following lemma is well-known and can be proved by induction, see [22, Lemma 8.3].

Lemma. *If $(h_1, \ldots, h_m)$ is a polycyclic sequence of G with relative orders $(r_1, \ldots, r_m)$, then every element $g \in G$ can be written uniquely as $g = h_1^{e_1} \ldots h_m^{e_m}$ with $e_i \in \mathbb{Z}$ for $1 \leq i \leq m$ and $0 \leq e_i < r_i$ if $r_i \neq \infty$.*

If $g = h_1^{e_1} \ldots h_m^{e_m}$ as in the lemma, then we refer to $h_1^{e_1} \ldots h_m^{e_m}$ as the **normal form** of g with respect to the polycyclic sequence $(h_1, \ldots, h_m)$.

4.1.1 Polycyclic presentations

We now describe polycyclic groups by certain group presentations. First, we recall the definition of a free group as we use it throughout the thesis.

Remark. For a set $\mathcal{G}$, we define $\mathcal{G}^{\pm} = \{(g, 1), (g, -1) \mid g \in \mathcal{G}\}$. We write g and g^{-1} for $(g, 1)$ and $(g, -1)$, respectively, and we define $(g^{-1})^{-1}$ to be equal to g. We assume that the symbol ε is not an element of $\mathcal{G}^{\pm}$. A **word in** $\mathcal{G}$ is either the empty word ε or it is a *string* $x_1 \ldots x_n$ where $n \in \mathbb{N}$ and $x_1, \ldots, x_n \in \mathcal{G}^{\pm}$. Two words u and v in $\mathcal{G}$ are **equivalent** if one can be obtained from the other by a finite number of insertions or deletions of substrings of the type xx^{-1} with $x \in \mathcal{G}^{\pm}$. This gives rise to an equivalence relation on the set of all words in $\mathcal{G}$, and we denote by $[u]$ the equivalence class of such a word u. The **free group $F_\mathcal{G}$ with free generating set $\mathcal{G}$** or the **free group on $\mathcal{G}$** is the set of these equivalence classes with multiplication $[u][v] = [uv]$, where uv is the concatenation of the strings u and v. Usually, we abuse notation and rather work with representatives of equivalence classes; that is, we consider the words in $\mathcal{G}$ up to equivalence and write $u = v$ whenever $[u] = [v]$. If $x \in \mathcal{G}^{\pm}$ and $e \geq 0$ is an integer, then we define x^e and x^{-e} to be equal to the concatenation of e copies of x and x^{-1}, respectively.

In general, a **group presentation** $\langle \mathcal{G} \mid \mathcal{R} \rangle$ consists of a set $\mathcal{G}$ and a subset $\mathcal{R}$ of the free group $F_\mathcal{G}$ with free generating set $\mathcal{G}$. The **group defined by the presentation** $\langle \mathcal{G} \mid \mathcal{R} \rangle$ is the quotient $G = F_\mathcal{G}/K$ where K is the normal closure of $\mathcal{R}$ in $F_\mathcal{G}$. We abuse notation and also write
$$G = \langle \mathcal{G} \mid \mathcal{R} \rangle.$$
The set $\mathcal{G}$ corresponds to a generating set of G and the elements of $\mathcal{R}$ are the **defining relators** of G. If $\mathcal{G}$ or $\mathcal{R}$ is finite, then we also write $\langle g_1, \ldots, g_n \mid \mathcal{R} \rangle$ and $\langle \mathcal{G} \mid r_1, \ldots, r_m \rangle$, respectively. Often, a relator $r \in \mathcal{R}$ is written as a **relation** $u = v$ where $r = u^{-1}v$. Two words u and w in $\mathcal{G}$ are **equivalent in** G if they represent the same element in G; that is, if $uK = wK$. We then write $u =_G w$ or say that $u = w$ in G.

Example. The group G defined by the presentation $\langle x, y \mid x^2 = y, y^2, [x, y] \rangle$ has the underlying set $\{1K, xK, yK, xyK\}$ where K is the normal subgroup of the free group on $\{x, y\}$ generated by the defining relators $\{x^{-2}y, y^2, x^{-1}y^{-1}xy\}$. Every word u in $\{x, y\}$ represents an element of G via the projection $u \mapsto uK$; for example, $y^2 x$ represents $y^2 xK = xyK$ and $y^2 x = xy$ in G.

We now consider polycyclic groups and recall the definition of a polycyclic presentation. For this purpose, let $\mathcal{G} = (g_1, \ldots, g_n)$ with $n \geq 1$ be an ordered set of abstract generators and let $(r_1, \ldots, r_n)$ be a list with $r_i \in \mathbb{N}$ or $r_i = \infty$. The set of indices $1 \leq i \leq n$ with $r_i \neq \infty$ is denoted by I. A word in the generators $\mathcal{G}$ is called **normalized** if it has the form $g_1^{e_1} \ldots g_n^{e_n}$ with each $e_i \in \mathbb{Z}$ and $0 \leq i < r_i$ whenever $i \in I$. Then, a **polycyclic presentation** with **exponents** $(r_1, \ldots, r_n)$ is a presentation of the form

$$\langle\ \mathcal{G}\ \mid\ g_i^{r_i} = w_{i,i}\ (i \in I),\quad g_k^{g_j} = w_{j,k}^+\ (j < k),\quad g_m^{g_l^{-1}} = w_{l,m}^-\ (l < m \text{ and } l \notin I)\ \rangle$$

where $w_{u,u}$, $w_{u,v}^-$, and $w_{u,v}^+$ are normalized words in $(g_{u+1}, \ldots, g_n)$. The relations of this presentation are called **power relations** and **conjugate relations**, respectively. Obviously, the group G defined by this presentation is polycyclic. The presentation is called **consistent** if every element in G can be represented by a unique normalized word in $\mathcal{G}$. This is the case if and only if the relative orders of the polycyclic sequence of G corresponding to $(g_1, \ldots, g_n)$ coincide with the exponents $(r_1, \ldots, r_n)$ of the presentation.

Conversely, it is straightforward to describe a given polycyclic group by a consistent polycyclic presentation with generating set corresponding to a polycyclic sequence.

A polycyclic presentation is also called a **power-conjugate presentation**, and we abbreviate the term "polycyclic presentation" by "**p.c.p.**".

4.1 Remark. Let $G = F/K$ be the group defined by the p.c.p. $\langle \mathcal{G} \mid \mathcal{R} \rangle$, where F is the free group on $\mathcal{G}$ and K is the normal closure of $\mathcal{R}$ in F. Let $\pi : F \to G$ be the projection,

4.1. Polycyclic groups

and choose a mapping (transversal) $\tau\colon G \to F$ with $\pi \circ \tau = \mathrm{id}_G$, whose image consists of normalized words in $\mathcal{G}$ and contains $1 \in F$. Note that there is exactly one such mapping if and only if the presentation is consistent. The set $\hat{G} = \{\tau(g) \mid g \in G\}$ with multiplication $\tau(g) \cdot \tau(h) = \tau(gh)$ is a group which is isomorphic to G via $\hat{G} \to G$, $\tau(h) \mapsto h$. The mapping $\varphi\colon F \to \hat{G}$, $u \mapsto \tau(\pi(u))$, is a homomorphism and, hence, every word u in $\mathcal{G}$ can be considered as an element of $\hat{G}$ via $u \mapsto \varphi(u)$.

We usually identify G with $\hat{G}$; that is, we assume that every element of G is represented by a fixed normalized word in $\mathcal{G}$. Note that $\hat{G}$ consists of *all* normalized words if and only if the presentation is consistent. Moreover, we abuse notation and consider every word u in $\mathcal{G}$ as an element of G via $u \mapsto \varphi(u)$; note that $u =_G \varphi(u)$ by construction.

We exemplify how we make use of this notation.

Example. Let A be the group defined by the consistent p.c.p. $\langle x, y \mid x^2 = y, y^2 = 1, y^x = y \rangle$. Following the conventions made in Remark 4.1, we identify A with the group of normalized words

$$\hat{A} = \{1, x, y, xy\}.$$

Let φ be the homomorphism which maps a word u in $\{x, y\}$ onto the unique element in $\hat{A}$ equivalent to u in A. Then the multiplication in $\hat{A}$ is the composition of φ and the usual concatenation of strings. For example, $x \cdot xy = \varphi(xxy) = \varphi(y^2) = 1$ and $y \cdot x = \varphi(yx) = xy$. We regard every word u in $\{x, y\}$ as an element of $\hat{A}$ via $u \mapsto \varphi(u)$. For example, we consider $xyx^2 \in A$ and $xxy \in A$ since $xyx^2 = x$ in A and $xxy =_A 1$.

Let B be the group defined by the consistent p.c.p. $\langle x, y \mid x^2 = 1, y^2 = 1, y^x = y \rangle$. By Remark 4.1, we identify

$$A = \{1, x, y, xy\} = B$$

as sets, and every word in $\{x, y\}$ can be considered as an element of A and B, respectively. For example, $xy^{-1}x^2 \in A$ as $xy^{-1}x^2 = x$ in A, and $xy^{-1}x^2 \in B$ as $xy^{-1}x^2 = xy$ in B, and we have to take care to minimize the risk of confusion.

4.1.2 Consistency

It is not obvious whether a given polycyclic presentation is consistent. In this paragraph, we recall a criterion for consistency as proved in [54].

Let G be the group defined by the p.c.p. $\langle \mathcal{G} \mid \mathcal{R} \rangle$ with generators $\mathcal{G} = (g_1, \ldots, g_n)$ and exponents $(r_1, \ldots, r_n)$, and let $I = \{i \mid 1 \leq i \leq n,\ r_i \neq \infty\}$. Every word w in $\mathcal{G}$ is equivalent in G to a normalized word w' and, in particular, the presentation is consistent if and only if the normalized word w' is unique for every w. A process of the determination of a normalized word which is equivalent to w in G is called a **collection** of w, and a *collection algorithm* is described in detail in [22, Section 8.1.3]. The general idea is to use the power and conjugate relations in $\mathcal{R}$ as *rewriting rules* to successively eliminate minimal non-normalized subwords of w, that is, subwords of the type $g_i g_j^{\pm 1}$ with $j < i$ or g_i^a with $i \in I$ and $a \notin \{0, \ldots, r_i - 1\}$. From now on, "collection" always refers to an application of this algorithm; we refer to [22] for more details and proofs.

The following theorem is proved in [54, p. 424], cf. [4, Lemma 2.1] and [16, Section 2.3]; we use the above notation.

4.2 Theorem. *The p.c.p. $\langle \mathcal{G} \mid \mathcal{R} \rangle$ is consistent if and only if for each of the following pairs of test words the collections of both words coincide, where the non-normalized subwords in brackets are collected first:*

$$\begin{aligned}
g_k(g_j g_i) &\quad \text{and} \quad (g_k g_j) g_i &&\text{for } k > j > i, \\
(g_j^{r_j}) g_i &\quad \text{and} \quad g_j^{r_j-1}(g_j g_i) &&\text{for } j > i,\ j \in I, \\
g_j(g_i^{r_i}) &\quad \text{and} \quad (g_j g_i) g_i^{r_i-1} &&\text{for } j > i,\ i \in I, \\
(g_i^{r_i}) g_i &\quad \text{and} \quad g_i(g_i^{r_i}) &&\text{for } i \in I, \\
g_j &\quad \text{and} \quad (g_j g_i^{-1}) g_i &&\text{for } j > i,\ i \notin I.
\end{aligned}$$

The application of Theorem 4.2 to a polycyclic presentation in order to test consistency is referred to as **doing consistency checks**.

Example. Let G be the group defined by the p.c.p.

$$\langle g_1, g_2, g_3 \mid g_1^4 = g_3,\ g_2^4 = g_3,\ g_3^4 = 1,\ g_2^{g_1} = g_2,\ g_3^{g_1} = g_3^2,\ g_3^{g_2} = g_3 \rangle.$$

The exponents of this presentation are $(4, 4, 4)$ and the normalized words in the generators are $\{g_1^{e_1} g_2^{e_2} g_3^{e_3} \mid 0 \leq e_1, e_2, e_3 \leq 3\}$. Consistency checks show that the presentation is not consistent. For example, the collections of $(g_3 g_1) g_1^3$ and $g_3(g_1^4)$ yield

$$(g_3 g_1) g_1^3 = g_1 g_3^2 g_1^3 = g_1 g_3 g_1 g_3^2 g_1^2 = g_1^2 g_3^4 g_1^2 = g_1^4 = g_3 \quad \text{and} \quad g_3(g_1^4) = g_3^2$$

in G. In particular, this shows that $g_3 =_G g_3^2$ and, thus, $g_3 = 1$ in G. One can deduce that G is isomorphic to the homocyclic group $C_4 \times C_4$ of rank 2.

4.2 Cohomology of polycyclic groups

We recall the connection between group extensions and second cohomology groups. Most of these results are standard and can be found in [22, 23, 47]. Afterwards, we use cohomology groups to describe the automorphism groups of certain group extensions. At the end of this section, we describe how to compute second cohomology groups for polycyclic groups. All these results are important tools in further investigations.

4.2.1 Group extensions and cohomology

Let N be a G-module. The group E is an **extension** of N by G if there exists a normal subgroup $M \trianglelefteq E$ such that $E/M \cong G$ and $M \cong N$. Using these isomorphisms, we usually identify N and G with M and E/M, respectively. If $\pi \colon E \to G$ is the projection, then (the image of) a mapping $\tau \colon G \to E$ with $\tau(1) = 1$ and $\pi \circ \tau = \mathrm{id}_G$ is a **transversal** to N in E. Via such a transversal, the extension E defines a mapping

$$\gamma \colon G \times G \to N, \quad (g, h) \mapsto \tau(gh)^{-1} \tau(g) \tau(h),$$

which for all $h, k, l \in G$ satisfies

$$\gamma(1, h) = \gamma(h, 1) = 1 \quad \text{and} \quad \gamma(h, k) \gamma(l, hk) = \gamma(lh, k) \gamma(l, h)^k.$$

4.2. Cohomology of polycyclic groups

Conversely, if δ is an element of

$$Z^2(G, N) = \{\gamma \colon G \times G \to N \mid \forall h, k, l \in G : \gamma(h, k)\gamma(l, hk) = \gamma(lh, k)\gamma(l, h)^k,$$
$$\gamma(1, h) = \gamma(h, 1) = 1\},$$

then the group $E(\delta)$ with underlying set $G \times N$ and multiplication

$$(g, n)(h, m) = (gh, n^h m \delta(g, h))$$

is an extension of N by G. If there is no risk of confusion, then we identify $G = G \times \{1\}$ and $N = \{1\} \times N$ and write gn for (g, n). By construction, if E is an extension of N by G defining a mapping $\gamma \colon G \times G \to N$ as above, then $E \cong E(\gamma)$.

The set $Z^2(G, N)$ carries the structure of an abelian group and is called the group of **2-cocycles** of G with coefficients in N. We write it additively even if N is written multiplicatively, that is, $(\gamma + \delta)(g, h) = \gamma(g, h)\delta(g, h)$ for all $\gamma, \delta \in Z^2(G, N)$ and $g, h \in G$.

An extension E of N by G is a **split extension** if there exists a transversal $\tau \colon G \to E$ which is a group homomorphism. In this case, E is isomorphic to the group $G \ltimes N$ with underlying set $G \times N$ and multiplication $(g, n)(h, m) = (gh, n^h m)$. If E is a split extension, then a corresponding 2-cocycle γ lies in

$$B^2(G, N) = \{\gamma \in Z^2(G, N) \mid \exists \delta \colon G \to N : \gamma(k, h) = \delta(kh)(\delta(k)^h)^{-1}\delta(h)^{-1}\},$$

which is the group of **2-coboundaries** of G with coefficients in N. In particular, the extension $E(\gamma)$ is a split extension if and only if $\gamma \in B^2(G, N)$.

Two extensions $E(\gamma)$ and $E(\delta)$ of N by G are **equivalent** if there exists an isomorphism φ from $E(\gamma)$ to $E(\delta)$ with $\varphi|_N = \mathrm{id}_N$ and $\varphi|_G = \mathrm{id}_G$, where the restriction to G is considered as the restriction to the quotient group modulo N. It is well-known that $E(\gamma)$ and $E(\delta)$ are equivalent if and only if $\gamma - \delta$ lies in $B^2(G, N)$. Thus, the **second cohomology group**

$$H^2(G, N) = Z^2(G, N) / B^2(G, N)$$

of G with coefficients in N describes all extensions of N by G up to equivalence. The elements of $H^2(G, N)$ are **cohomology classes** and for a cohomology class $\delta + B^2(G, N)$ we define the extension $E(\delta + B^2(G, N))$ as $E(\delta)$. Note that this definition depends on the chosen coset representative δ, but for all choices we obtain equivalent extensions.

Two extensions $E(\gamma)$ and $E(\delta)$ of N by G are **strongly isomorphic** if there exists an isomorphism from $E(\gamma)$ to $E(\delta)$ which maps $N \leq E(\gamma)$ onto $N \leq E(\delta)$. We now describe a construction of group extensions up to strong isomorphism.

If $\overline{g} \in \mathrm{Aut}(N)$ denotes the action of $g \in G$ on N, then the **group of compatible pairs** of G and N is defined as

$$\mathrm{Comp}(G, N) = \{(\alpha, \beta) \in \mathrm{Aut}(G) \times \mathrm{Aut}(N) \mid \forall g \in G : \beta \circ \overline{g} \circ \beta^{-1} = \overline{\alpha(g)}\}.$$

Thus, if $(\alpha, \beta) \in \mathrm{Comp}(G, N)$, then $\beta(n)^{\alpha(g)} = \beta(n^g)$ for all $g \in G$ and $n \in N$. It is straightforward, but technical, to prove that a compatible pair $(\alpha, \beta) \in \mathrm{Comp}(G, N)$ acts on $Z^2(G, N)$ via

$$\gamma \mapsto \gamma^{(\alpha, \beta)} = \beta^{-1} \circ \gamma \circ (\alpha, \alpha),$$

and $B^2(G, N)$ is invariant under this action. Therefore, the group of compatible pairs acts on $H^2(G, N)$, cf. [1, Section 4.2.1] and [22, p. 55]. The following theorem shows that the orbits under this action correspond to extensions up to strong isomorphism.

4.3 Theorem. *Let N be a G-module and let $\gamma, \delta \in Z^2(G, N)$ be 2-cocycles. The extensions $E(\gamma)$ and $E(\delta)$ are strongly isomorphic if and only if $\delta^c - \gamma$ lies in $B^2(G, N)$ for some compatible pair $c \in \mathrm{Comp}(G, N)$.*

Proof. We sketch the proof as given in [1, Theorem 4.7]. First, if $\varphi \colon E(\gamma) \to E(\delta)$ is a strong isomorphism, then $(\alpha, \beta) = (\varphi|_G, \varphi|_N)$ is a compatible pair of G and N. Let $\varepsilon \colon G \to N$ be defined by $\varphi((\alpha^{-1}(g), 1)) = (g, \varepsilon(g))$. If $u = (\alpha^{-1}(g), 1)$ and $v = (\alpha^{-1}(h), 1)$ with $g, h \in G$, then it follows from $\varphi(uv) = \varphi(u)\varphi(v)$ that

$$\delta(g, h) = \gamma^{(\alpha^{-1}, \beta^{-1})}(g, h)\varepsilon(gh)(\varepsilon(g)^h)^{-1}\varepsilon(h)^{-1},$$

which shows that $\delta^{(\alpha, \beta)} - \gamma \in B^2(G, N)$. Conversely, if (α, β) is a compatible pair with $\delta^{(\alpha, \beta)} = \gamma + \psi^{(\alpha, \beta)}$ for some 2-coboundary ψ defined by $\varepsilon \colon G \to N$, then

$$E(\gamma) \to E(\delta), \quad (g, n) \mapsto (\alpha(g), \varepsilon(\alpha(g))\beta(n)),$$

is a strong isomorphism. □

The following example shows that in general a reduction of group extensions up to strong isomorphism is *not* a reduction up to isomorphism.

Example. Let G be the group of order 32 defined by the consistent p.c.p.

$$\langle\; a_1, \ldots, a_5 \;\mid\; a_1^2 = 1,\; a_2^2 = a_4,\; a_3^2 = 1,\; a_4^2 = 1,\; a_5^2 = 1,\; a_2^{a_1} = a_2 a_4,$$
$$a_3^{a_1} = a_3 a_5,\; a_4^{a_1} = a_4,\; a_5^{a_1} = a_5,\; a_i^{a_j} = a_i \;(2 \leq j < i)\;\rangle.$$

The center of G is generated by $\{a_4, a_5\}$ and we denote by N and M the cyclic subgroups of G generated by a_4 and a_5, respectively. One can prove that G/N and G/M are isomorphic, and N and M are characteristic in G. This shows that G can be considered as an extension by M and by N, respectively, and these two extensions are isomorphic but not strongly isomorphic.

4.2.2 Automorphism groups of extensions

For a group G and a G-module N the corresponding groups of **1-cocycles**, **1-coboundaries**, and the **first cohomology group** are defined as follows.

$$\begin{aligned}
Z^1(G, N) &= \{\gamma \colon G \to N \mid \forall g, h \in G : \gamma(gh) = \gamma(g)^h \gamma(h)\}, \\
B^1(G, N) &= \{\gamma \in Z^1(G, N) \mid \exists n \in N \; \forall g \in G : \gamma(g) = n^g n^{-1}\}, \\
H^1(G, N) &= Z^1(G, N)/B^1(G, N).
\end{aligned}$$

Analogously to the group of 2-cocycles, the group of compatible pairs $\mathrm{Comp}(G, N)$ acts on $Z^1(G, N)$ and $H^1(G, N)$, respectively.

We now use cohomology to describe the automorphism groups of certain extensions of N by G. For this purpose, let E be an extension of N by G. We assume that N is characteristic in E and we may assume that $E = E(\gamma)$ for some 2-cocycle $\gamma \in Z^2(G, N)$. Again, let $\overline{g} \in \mathrm{Aut}(N)$ be the action of $g \in G$ on N. In the following lemma, we consider the homomorphism

$$\phi \colon \mathrm{Aut}(E) \to \mathrm{Aut}(G) \times \mathrm{Aut}(N), \quad \alpha \mapsto (\alpha|_G, \alpha|_N),$$

and determine its kernel and image, see [22, Section 8.9].

4.2. Cohomology of polycyclic groups

4.4 Lemma. $\ker \phi \cong Z^1(G, N)$ *and* $\operatorname{im} \phi = \operatorname{Stab}_{\operatorname{Comp}(G,N)}(\gamma + B^2(G, N))$.

Proof. If $\alpha \in \ker \phi$, then $\alpha \colon E \to E$, $(g, n) \mapsto (g, n\psi_\alpha(g))$, for some $\psi_\alpha \in Z^1(G, N)$. Conversely, if $\psi \in Z^1(G, N)$ is given, then $(g, n) \mapsto (g, n\psi(g))$ is an automorphism of E lying in $\ker \phi$. The mapping $\ker \phi \to Z^1(G, N)$, $\alpha \mapsto \psi_\alpha$, is a group isomorphism.

It follows from the proof of Theorem 4.3 that $\operatorname{im} \phi = \operatorname{Stab}_{\operatorname{Comp}(G,N)}(\gamma + B^2(G, N))$. $\square$

4.2.3 Second cohomology of polycyclic groups

Throughout this section, let G be a polycyclic group and let A be a G-module. We assume that G and A are the groups defined by the consistent p.c.p.s $\langle \mathcal{G} \mid \mathcal{R} \rangle$ and $\langle \mathcal{A} \mid \mathcal{B} \rangle$, respectively, with disjoint generating sets $\mathcal{G} = (g_1, \ldots, g_n)$ and $\mathcal{A} = (a_1, \ldots, a_m)$. By definition, G and A consist of normalized words in $\mathcal{G}$ and $\mathcal{A}$, respectively. We now describe a method to compute the second cohomology group $H^2(G, A)$ by applying consistency checks to a polycyclic presentation containing certain indeterminates. The results of this section are partly from and motivated by [4, Section 2.3] and [22, Section 8.7.2].

As a first step, we determine a consistent p.c.p. of the extension $E(\gamma)$ defined by a 2-cocycle $\gamma \colon G \times G \to A$. Recall that $E(\gamma)$ has the underlying set $G \times A$ and multiplication $(f, a)(g, b) = (fg, a^g b \gamma(f, g))$, and we identify (f, a) with fa for $f \in G$ and $a \in A$. Let $\tau \colon G \to E(\gamma)$ be the transversal which maps a normalized word $g \in G$ onto $\tau(g) = g$. Via this transversal, every relation r in $\mathcal{R}$ can be evaluated in $E(\gamma)$. If r is a power relation saying $g_i^{r_i} = w_{i,i}$, then there is a **tail** $x_r \in A$ such that $g_i^{r_i} = w_{i,i} x_r$ in $E(\gamma)$; we denote this relation by $r.x_r$. Analogously, we define $r.x_r$ for a conjugate relation r in $\mathcal{R}$. All these tails are collected in a list $x_\gamma = (x_r)_{r \in \mathcal{R}}$ which is called the **tail vector** defined by γ. Finally, let $\mathcal{C}$ be the set of conjugate relations describing the G-action on A; that is, the elements of $\mathcal{C}$ are of the type $a^g = w_{a,g}$ with $a \in \mathcal{A}$, $g \in \mathcal{G}$, and $w_{a,g} \in A$. We now show how x_γ can be used to define a consistent p.c.p. of $E(\gamma)$ on the generating set $\mathcal{G} \cup \mathcal{A}$.

Definition. Let $x = (x_r)_{r \in \mathcal{R}}$ be a list of elements in A. Then $\mathcal{E}(x)$ is the polycyclic presentation with generating set $\mathcal{G} \cup \mathcal{A}$ and defining relations $\{r.x_r \mid r \in \mathcal{R}\} \cup \mathcal{B} \cup \mathcal{C}$. The group defined by $\mathcal{E}(x)$ is denoted by $E(x)$.

4.5 Example. We consider the groups

$$G = \langle g_1, g_2 \mid g_1^2 = 1,\ g_2^2 = 1,\ g_2^{g_1} = g_2 \rangle \quad \text{and} \quad A = \langle a_1 \mid a_1^4 = 1 \rangle.$$

Let $\mathcal{C} = \{a_1^{g_1} = a_1^3,\ a_1^{g_2} = a_1\}$ and, corresponding to the relations $(g_1^2 = 1,\ g_2^2 = 1,\ g_2^{g_1} = g_2)$, we consider a list $x = (a_1^2, a_1^3, a_1^3)$ of words in $\{a_1\}$. Then the p.c.p. $\mathcal{E}(x)$ is defined as

$$\mathcal{E}(x) = \langle\ g_1, g_2, a_1 \mid g_1^2 = a_1^2,\ g_2^2 = a_1^3,\ a_1^4 = 1,\ g_2^{g_1} = g_2 a_1^3,\ a_1^{g_1} = a_1^3,\ a_1^{g_2} = a_1\ \rangle,$$

and one can show that $E(x)$ has order 16.

4.6 Lemma. a) *If $x = (x_r)_{r \in \mathcal{R}}$ is a list of elements in A, then $E(x)$ is an extension of A by G if and only if the presentation $\mathcal{E}(x)$ is consistent.*

b) *If $\gamma \in Z^2(G, A)$, then the presentation $\mathcal{E}(x_\gamma)$ is consistent and the extensions $E(x_\gamma)$ and $E(\gamma)$ are equivalent.*

Proof. a) The quotient of $E(x)$ modulo the normal subgroup $\tilde{A}$ generated by $\mathcal{A}$ is isomorphic to G. Thus, the group $E(x)$ is an extension of A by G if and only if $A \cong \tilde{A}$, if and only if every element in $E(x)$ is represented by a unique normalized word in $\mathcal{G} \cup \mathcal{A}$, if and only if $\mathcal{E}(x)$ is consistent.

b) The sequence $(g_1, \ldots, g_n, a_1, \ldots, a_m)$ is a polycyclic sequence of $E(\gamma)$ and, by construction, the corresponding consistent p.c.p. with generating set $\mathcal{G} \cup \mathcal{A}$ is $\mathcal{E}(x_\gamma)$. □

Lemma 4.6 shows that every extension of A by G can be described by an element of
$$A^\mathcal{R} = \{(x_r)_{r \in \mathcal{R}} \mid x_r \in A \text{ for all } r \in \mathcal{R}\}.$$

Note that $A^\mathcal{R}$ is an abelian group with addition $(x_r)_{r \in \mathcal{R}} + (y_r)_{r \in \mathcal{R}} = (x_r y_r)_{r \in \mathcal{R}}$.

Definition. Using the above construction, every 2-cocycle γ defines a unique tail vector x_γ, which yields a mapping
$$\phi \colon Z^2(G, A) \to A^\mathcal{R}, \quad \gamma \mapsto x_\gamma,$$
and we define
$$\mathcal{Z}(G, A) = \phi(Z^2(G, A)) \quad \text{and} \quad \mathcal{B}(G, A) = \phi(B^2(G, A)).$$

Obviously, this definition depends on the chosen presentations of G and A.

As in [22, Lemma 8.47], we now prove that ϕ is a group homomorphism. This allows us to call $\mathcal{Z}(G, A)$ and $\mathcal{B}(G, A)$ the **group of tail vectors** and **group of coboundary tail vectors** of G in A, respectively.

4.7 Lemma. *The mapping ϕ is a group homomorphism with $\ker \phi \leq B^2(G, A)$, and so*
$$H^2(G, A) \cong \mathcal{Z}(G, A) / \mathcal{B}(G, A).$$

Proof. If $\gamma_1, \gamma_2 \in Z^2(G, A)$, then $(\gamma_1 + \gamma_2)(g, h) = \gamma_1(g, h) \gamma_2(g, h)$ for all $g, h \in G$. This implies that $\phi(\gamma_1 + \gamma_2) = \phi(\gamma_1) + \phi(\gamma_2)$ and ϕ is a homomorphism. If $\gamma \in \ker \phi$, then $E(x_\gamma)$ is a split extension, which shows that $\gamma \in B^2(G, A)$ as $E(\gamma)$ and $E(x_\gamma)$ are equivalent. Thus, ϕ induces a homomorphism $Z^2(G, A) \to \mathcal{Z}(G, A) / \mathcal{B}(G, A)$ with kernel $B^2(G, A)$. □

Again, the following definition depends on the chosen presentations of G and A.

4.8 Definition. If $x \in A^\mathcal{R}$ defines a consistent p.c.p. $\mathcal{E}(x)$, then the canonical 2-cocycle defined by x is
$$\gamma_x \colon G \times G \to A, \quad (h, g) \mapsto \tau(gh)^{-1} \tau(g) \tau(h),$$
where $\tau \colon G \to E(x)$ is the transversal which maps a normalized word g onto $\tau(g) = g$.

4.9 Lemma. *If $x \in A^\mathcal{R}$ defines a consistent p.c.p. $\mathcal{E}(x)$, then $\phi(\gamma_x) = x$.*

Proof. By definition, the value of $\gamma_x(g, h)$ with $g, h \in G$ corresponds to the tail in A occurring at the collection of the product of $\tau(g)$ and $\tau(h)$ in $E(x)$. We use this property to prove the assertion. Let $\hat{\tau} \colon G \to E(\gamma_x)$ be the transversal which maps a normalized word $g \in G$ onto $\hat{\tau}(g) = g$, and denote by $t = (t_r)_{r \in \mathcal{R}}$ the tail vector defined by γ_x. If $r \in \mathcal{R}$ is the power-relation $g_i^{r_i} = w_{i,i}$, then $E(\gamma_x)$ satisfies $\hat{\tau}(g_i)^{r_i} = \hat{\tau}(w_{i,i}) t_r$ and
$$\hat{\tau}(g_i)^{r_i} = \hat{\tau}(w_{i,i}) \prod_{k=1}^{r_i - 1} \gamma_x(g_i, g_i^k) = \hat{\tau}(w_{i,i}) \gamma_x(g_i, g_i^{r_i - 1}) = \hat{\tau}(w_{i,i}) x_r.$$

4.2. Cohomology of polycyclic groups

If r is the conjugate relation $g_i^{g_j} = w_{j,i}^+$, then $E(\gamma_x)$ satisfies $\hat{\tau}(g_i)^{\hat{\tau}(g_j)} = \hat{\tau}(w_{j,i}^+)t_r$ and

$$\hat{\tau}(g_i)\hat{\tau}(g_j) = \hat{\tau}(g_j)\hat{\tau}(w_{j,i}^+)\gamma_x(g_i,g_j)\gamma_x(g_j,w_{j,i}^+)^{-1} = \hat{\tau}(g_j)\hat{\tau}(w_{j,i}^+)x_r.$$

A similar computation shows that $t_r = x_r$ if r is the conjugate relation $g_i^{g_j^{-1}} = w_{j,i}^-$; that is, $t_r = x_r$ for all $r \in \mathcal{R}$. This proves that $\phi(\gamma_x) = x$. $\square$

The following corollary is a consequence of Lemmas 4.6 and 4.9.

4.10 Corollary. *If $x \in A^{\mathcal{R}}$, then the following are equivalent:*
(1) The presentation $\mathcal{E}(x)$ is consistent.
(2) The group $E(x)$ is an extension of A by G.
(3) The list x is a tail vector of G in A, that is, $x \in \mathcal{Z}(G,A)$.

Thus, the tail vectors in $\mathcal{Z}(G,A)$ are exactly those elements of $A^{\mathcal{R}}$ which define consistent presentations. By Theorem 4.2, these can be determined by applying consistency checks to a presentation $\mathcal{E}(x)$ where $x = (x_r)_{r \in \mathcal{R}}$ is a list of indeterminate elements $x_r \in A$.

We now describe the computation of the group of coboundary tail vectors. By construction, a tail vector x lies in $\mathcal{B}(G,A)$ if and only if $E(x)$ is a split extension of A by G, if and only if the group A has a complement in $E(x)$ which is generated by $\{ga_g \mid g \in \mathcal{G}\}$ for some $a_g \in A$. Such a generating set has to fulfill the relations in $\mathcal{R}$, which imposes necessary and sufficient conditions on x to be a coboundary tail vector. This allows us to determine the elements of $\mathcal{B}(G,A)$.

As an easy example, we compute the second cohomology group of the elementary abelian group of order p^2 with coefficients in the cyclic group of order p.

Example. We consider the group $G = \langle \mathcal{G} \mid \mathcal{R} \rangle$ and the trivial G-module $A = \langle \mathcal{A} \mid \mathcal{B} \rangle$ with generating sets $\mathcal{G} = \{g_1, g_2\}$ and $\mathcal{A} = \{a_1\}$ and defining relations

$$\mathcal{R} = \{g_1^p = 1,\ g_2^p = 1,\ g_2^{g_1} = g_2\} \quad \text{and} \quad \mathcal{B} = \{a_1^p = 1\}.$$

For a list $x = (x_1, x_2, x_3)$ of elements in A, the presentation $\mathcal{E}(x)$ is defined as

$$\langle g_1, g_2, a_1 \mid g_1^p = x_1,\ g_2^p = x_2,\ g_2^{g_1} = g_2 x_3,\ a_1^p = 1,\ a_1^{g_1} = a_1,\ a_1^{g_2} = a_1 \rangle.$$

In order to compute $\mathcal{Z}(G,A)$, we consider $x = (x_1, x_2, x_3)$ as a list of indeterminates in A and apply consistency checks to the presentation $\mathcal{E}(x)$, see Theorem 4.2. A straightforward computation shows that $\mathcal{E}(x)$ is consistent for all choices of x_1, x_2, and x_3. For example, the collections of $a_1(g_2 g_1)$ and $(a_1 g_2) g_1$ yield

$$a_1(g_2 g_1) = a_1 g_1 g_2 x_3 = g_1 a_1 g_2 x_3 = g_1 g_2 a_1 x_3 \text{ in } E(x)$$

and

$$(a_1 g_2) g_1 = g_2 a_1 g_1 = g_2 g_1 a_1 = g_1 g_2 x_3 a_1 \text{ in } E(x),$$

that is, they impose no condition on the tail x_3. Thus, it is shown that

$$\mathcal{Z}(G,A) = A^{\mathcal{R}} \cong C_p^3.$$

We now determine $\mathcal{B}(G,A)$ and consider a generating set $\{g_1u, g_2v\}$ of a complement to A in $E(x)$. The elements g_1u and g_2v have to fulfill the relations of G and, therefore,

$$1 = (g_1u)^p = g_1^p u^p = x_1 \text{ in } E(x)$$

and, analogously, $x_2 =_{E(x)} 1$. It follows from $g_2v = (g_2v)^{g_1u} = g_2 x_3 v$ in $E(x)$ that $x_3 =_{E(x)} 1$, and $\mathcal{B}(G,A) = \{(1,1,1)\}$ is trivial. This shows that $H^2(G,A) \cong C_p^3$.

5 Number theory

The aim of this chapter is to summarize the number theory used in this thesis. In the first part of this chapter, we recall some p-adic number theory which can be found in [18, 19, 40]. Afterwards, based on [7], we briefly consider pro-p groups and recall their $\mathbb{Z}_p$-group structure. Motivated by the definition of skeleton groups, see Definition 2.2, we then examine the group of P-homomorphisms $T \wedge T \to T$ in the second part; the results provided there are based on [29, Section 8.3]. An action on this group of homomorphisms is defined in the third part, which is motivated by [28, 29]. This action plays an important role in the construction of skeleton groups up to isomorphism.

5.1 The p-th local cyclotomic field

The results of this section are standard and can be found in [18, 19, 40].

5.1.1 p-adic numbers

The p-adic valuation $\nu_p \colon \mathbb{Q} \to \mathbb{Z} \cup \{\infty\}$ is uniquely defined by $\nu_p(0) = \infty$ and, for $x \neq 0$, by the formula
$$x = p^{\nu_p(x)} \tfrac{a}{b}, \quad p \nmid ab.$$
The p-adic absolute value $|\cdot|_p$ on $\mathbb{Q}$ is defined by $|x|_p = p^{-\nu_p(x)}$ if $x \neq 0$ and $|0|_p = 0$. It is **non-archimedean**, that is, it satisfies $|x + y|_p \leq \max\{|x|_p, |y|_p\}$ for all $x, y \in \mathbb{Q}$. The field $\mathbb{Q}$ is not complete with respect to this absolute value and its completion is the field of p-**adic numbers** $\mathbb{Q}_p$. By definition, $\mathbb{Q}_p$ consists of equivalence classes of Cauchy sequences. The p-adic absolute value can be extended to $\mathbb{Q}_p$ via $|x|_p = \lim_{n \to \infty} |x_n|_p$ where $(x_n)_{n \in \mathbb{N}}$ is a Cauchy sequence in $\mathbb{Q}_p$ representing x.

The ring of p-adic integers $\mathbb{Z}_p$ is the **valuation ring** $\{x \in \mathbb{Q}_p \mid |x|_p \leq 1\}$ and can be considered as
$$\mathbb{Z}_p = \left\{ \sum_{i=0}^{\infty} a_i p^i \mid \forall i \colon 0 \leq a_i \leq p-1,\ a_i \in \mathbb{N} \right\}.$$
The group of p-**adic units** $\mathbb{Z}_p^\star = \{x \in \mathbb{Q}_p \mid |x|_p = 1\}$ consists of the elements in $\mathbb{Z}_p$ which are invertible in $\mathbb{Z}_p$.

A well-known lemma is the following, see [18, Theorem 3.4.6]. We denote by $\overline{f} \in \mathbb{F}_p[X]$ the polynomial defined by $f \in \mathbb{Z}_p[X]$ via a coefficient-wise reduction modulo p.

Hensel's Lemma. *Let $f \in \mathbb{Z}_p[X]$ and assume that there exist $g_0, g_1 \in \mathbb{Z}_p[X]$ such that g_0 is monic, $\gcd(\overline{g_0}, \overline{g_1}) = 1$ in $\mathbb{F}_p[X]$, and $f \equiv g_0 g_1 \bmod p$ (coefficient-wise). Then there exist $h_0, h_1 \in \mathbb{Z}_p[X]$ such that h_0 is monic, $g_i \equiv h_i \bmod p$ for $i = 1, 2$, and $f = h_0 h_1$.*

It can be deduced that the p-adic integers contain a primitive $(p-1)$-th root of unity $\omega \in \mathbb{Z}_p$. If $1 < r < p$ represents a generator of the multiplicative group $\mathbb{F}_p^\star$, then ω can be defined as $\omega = \lim_{n \to \infty} r^{(p^n)}$ so that $\omega \equiv r^{(p^{n-1})} \bmod p^n$ for all positive integers n.

5.1 Remark. Let $\overline{\mathbb{Q}_p}$ be the algebraic closure of $\mathbb{Q}_p$, see [18, p. 153] and let $\mathbb{C}_p$ be the completion of $\overline{\mathbb{Q}_p}$, see [18, Proposition 5.7.6]. Both fields are algebraically closed and both have transcendence degree equal to the cardinality of the reals. The same holds for the field $\mathbb{C}$ of the complex numbers. Now a theorem of Steinitz proves that $\overline{\mathbb{Q}_p} \cong \mathbb{C} \cong \mathbb{C}_p$ as fields, see [46, p. 145] for a proof. Thus, there is an embedding from $\mathbb{Q}_p$ into $\mathbb{C}$ and every p-adic root of unity $u \in \mathbb{Q}_p$ can be considered as a complex root of unity $u \in \mathbb{C}$.

5.1.2 Cyclotomic fields

Let θ be a primitive p-th root of unity over $\mathbb{Q}_p$. The **p-th local cyclotomic field** is defined as $\mathbb{Q}_p(\theta)$; that is,
$$\mathbb{Q}_p(\theta) \cong \mathbb{Q}_p[X]/(1 + X + \ldots + X^{p-1})\mathbb{Q}_p[X]$$
and $\mathbb{Q}_p(\theta)$ has $\mathbb{Q}_p$-basis $\{1, \theta, \ldots, \theta^{p-2}\}$, see [18, Section 5.6].

The p-adic absolute value $|\cdot|_p$ on $\mathbb{Q}_p$ extends to $\mathbb{Q}_p(\theta)$ and, with respect to this absolute value, $\mathbb{Q}_p(\theta)$ is complete. The **valuation ring** of $\mathbb{Q}_p(\theta)$ is $\{x \in \mathbb{Q}_p(\theta) \mid |x|_p \leq 1\}$ and coincides with
$$\mathbb{Z}_p[\theta] = \left\{ a_0 + a_1\theta + \ldots + a_{p-2}\theta^{p-2} \mid a_0, \ldots, a_{p-2} \in \mathbb{Z}_p \right\}.$$

The invertible elements in $\mathbb{Z}_p[\theta]$ are $\{x \in \mathbb{Q}_p(\theta) \mid |x|_p = 1\}$ and we denote this **group of units** by
$$\mathcal{U}_p = \mathbb{Z}_p[\theta]^\star.$$
The element
$$\kappa = \theta - 1$$
is a **prime element** of $\mathbb{Q}_p(\theta)$; that is, every $x \in \mathbb{Q}_p(\theta) \setminus \{0\}$ can be written uniquely as $x = \kappa^z u$ for some integer z and a unit $u \in \mathcal{U}_p$.

The ring $\mathbb{Z}_p[\theta]$ is a principal domain with unique maximal ideal
$$\mathfrak{p} = \{x \in \mathbb{Q}_p(\theta) \mid |x|_p < 1\}$$
generated by κ. For $z \in \mathbb{Z}$ we define
$$\mathfrak{p}^z = \{\kappa^z t \mid t \in \mathbb{Z}_p[\theta]\},$$
and the non-zero ideals of $\mathbb{Z}_p[\theta]$ are $\mathfrak{p}^0, \mathfrak{p}^1, \mathfrak{p}^2, \ldots$ The ideal $\mathfrak{p}^{p-1}$ is generated by p and, thus, if $n = x(p-1) + i$ with integers $x \geq 0$ and $0 \leq i \leq p-2$, then $\mathfrak{p}^n = p^x \mathfrak{p}^i$. The factor group $(\mathfrak{p}^n/\mathfrak{p}^{n+1}, +)$ is cyclic of order p and the **residue class field** $\mathbb{Z}_p[\theta]/\mathfrak{p}$ of $\mathbb{Q}_p(\theta)$ is isomorphic to $\mathbb{F}_p$.

The Galois group
$$\mathcal{G}(\mathbb{Q}_p(\theta)/\mathbb{Q}_p) = \{\alpha \in \mathrm{Aut}(\mathbb{Q}_p(\theta)) \mid \forall x \in \mathbb{Q}_p : \alpha(x) = x\}$$
is cyclic of order $p-1$ and generated by an automorphism defined by $\theta \mapsto \theta^h$ for some $1 < h < p$ generating $\mathbb{F}_p^\star$, see [40, Satz II.7.13].

5.1.3 Group of units

We now determine the structure of the group of units $\mathcal{U}_p$. The following theorem is proved in [40, Satz II.5.3]; recall that $\omega \in \mathbb{Z}_p$ is a primitive $(p-1)$-th root of unity.

Theorem. *The group of units $\mathcal{U}_p$ can be written as $\mathcal{U}_p = C_{p-1}(\omega) \times (1 + \mathfrak{p})$.*

This direct product is considered as an internal direct product; that is, every $u \in \mathcal{U}_p$ can be written uniquely as $u = \omega^a(1+t)$ with $0 \le a \le p-2$ and $t \in \mathfrak{p}$.

The multiplicative group $\mathcal{U}_p^{(1)} = 1 + \mathfrak{p}$ is the group of **one-units** and, more general, the group of i-th one-units with $i \ge 1$ is defined as

$$\mathcal{U}_p^{(i)} = 1 + \mathfrak{p}^i.$$

According to [19, Section 15.7], we can write

$$\mathcal{U}_p = C_{p-1}(\omega) \times C_p(\theta) \times \mathcal{U}_p^{(2)}.$$

The group of i-th one-units has the structure of a $\mathbb{Z}_p$-module where $z \in \mathbb{Z}_p$ acts on $u \in \mathcal{U}_p^{(i)}$ via $u^z = \lim_{n \to \infty} u^{z_n}$ where $z_n = z \bmod p^n$ for $n \ge 1$. In particular, the group of second one-units is a **free $\mathbb{Z}_p$-module of rank** $p-1$, that is, it is isomorphic to $\mathbb{Z}_p^{p-1}$, see [19, Section 15.7].

Theorem. *As a $\mathbb{Z}_p$-module, the group $\mathcal{U}_p^{(2)}$ is freely generated by $\{1-(1-\theta)^h \mid 2 \le h \le p\}$.*

In particular, the multiplicative group of i-th one-units and the additive group $(\mathfrak{p}^i, +)$ are isomorphic if $i \ge 2$; for a proof we refer to [40, Satz II.5.5 & p. 146].

5.2 Theorem. *For $i \ge 2$ the power series*

$$\exp(x) = 1 + x + \frac{x^2}{2!} + \frac{x^3}{3!} + \ldots \quad \text{and} \quad \log(1+x) = x - \frac{x^2}{2} + \frac{x^3}{3} - \ldots$$

define isomorphisms $\exp\colon \mathfrak{p}^i \to \mathcal{U}_p^{(i)}$ *and* $\log\colon \mathcal{U}_p^{(i)} \to \mathfrak{p}^i$ *of $\mathbb{Z}_p$-modules with $\exp^{-1} = \log$.*

5.1.4 The unit group as a centralizer

If R is a commutative ring, then we denote by $M(m, R)$ the R-module of all $m \times m$ matrices over R, and the centralizer $C_{M(m,R)}(M)$ of a matrix $M \in M(m, R)$ is defined as the additive group

$$C_{M(m,R)}(M) = \{N \in M(m, R) \mid NM = MN\}.$$

Recall that $d = p - 1$, see Definition 2.1, and we denote by

$$\mathfrak{G} = \begin{pmatrix} 0 & 1 & 0 & \cdots & 0 \\ 0 & 0 & 1 & \cdots & 0 \\ \vdots & & & \ddots & \\ 0 & 0 & 0 & \cdots & 1 \\ -1 & -1 & -1 & \cdots & -1 \end{pmatrix} \in \mathrm{GL}(d, \mathbb{Z}_p)$$

the companion matrix of $1 + X + \ldots + X^d$ over $\mathbb{Z}_p$.

5.3 Lemma. a) If $R \in \{\mathbb{Z}_p, \mathbb{Q}_p\}$, then $C_{M(d,R)}(\mathfrak{G})$ is a ring and

$$C_{M(d,R)}(\mathfrak{G}) = \left\{ \sum_{i=0}^{d-1} a_i \mathfrak{G}^i \mid \forall i : a_i \in R \right\}.$$

b) An isomorphism $C_{\mathrm{GL}(d,\mathbb{Z}_p)}(\langle\mathfrak{G}\rangle) \cong \mathcal{U}_p$ is induced by $\sum_{i=0}^{d-1} a_i \mathfrak{G}^i \mapsto \sum_{i=0}^{d-1} a_i \theta^i$.

c) There exists $\mu \in \mathrm{GL}(d,\mathbb{Z}_p)$ of order d such that $\mathfrak{G}^\mu = \mathfrak{G}^j$ for some $1 < j < d$ and

$$N_{\mathrm{GL}(d,\mathbb{Z}_p)}(\langle\mathfrak{G}\rangle) = \left\{ \mu^i u \mid 0 \leq i \leq d,\ u \in C_{\mathrm{GL}(d,\mathbb{Z}_p)}(\langle\mathfrak{G}\rangle) \right\}.$$

Proof. For a positive integer m let I_m be the $m \times m$ identity matrix.
a) If $M \in C_{M(d,R)}(\mathfrak{G})$, then it follows from $M\mathfrak{G} = \mathfrak{G}M$ that M is determined uniquely by the entries of its first row. By induction, the first row of $\mathfrak{G}^i$ is the $(i+1)$-th row of I_d for $0 \leq i \leq d-1$ and, thus, M can be written as an R-linear combination of $\{\mathfrak{G}^0, \ldots, \mathfrak{G}^{d-1}\}$.
b) By definition, $\mathfrak{G}^0 + \mathfrak{G}^1 + \ldots + \mathfrak{G}^d = 0 I_d$ and, by part a), there is a field isomorphism $C_{M(d,\mathbb{Q}_p)}(\mathfrak{G}) \cong \mathbb{Q}_p(\theta)$; that is, $C_{M(d,\mathbb{Z}_p)}(\mathfrak{G}) \cong \mathbb{Z}_p[\theta]$ as rings and $C_{\mathrm{GL}(d,\mathbb{Z}_p)}(\langle\mathfrak{G}\rangle) \cong \mathcal{U}_p$.
c) The group $C = C_{\mathrm{GL}(d,\mathbb{Z}_p)}(\langle\mathfrak{G}\rangle)$ is a normal subgroup of $N = N_{\mathrm{GL}(d,\mathbb{Z}_p)}(\langle\mathfrak{G}\rangle)$ and the quotient N/C embeds into $\mathrm{Aut}(\langle\mathfrak{G}\rangle) \cong C_d$. We choose μ as the matrix describing the action of a generator of $\mathcal{G}(\mathbb{Q}_p(\theta)/\mathbb{Q}_p)$ on $\mathbb{Z}_p[\theta]$ with respect to $\{1, \theta, \ldots, \theta^{d-1}\}$. Then μ lies in $N \setminus C$, which proves the assertion. $\square$

5.2 Pro-p groups and $\mathbb{Z}_p$-modules

We briefly consider pro-p groups and recall their $\mathbb{Z}_p$-operator group structure. For proofs and background we refer to [7, Chapter 1].

A **directed set** is a non-empty partially ordered set $(\Lambda, \geq)$ with the property that for every $\lambda, \mu \in \Lambda$ there exists $\nu \in \Lambda$ with $\nu \geq \lambda$ and $\nu \geq \mu$. An **inverse system** of groups over Λ is a family of groups $(G_\lambda)_{\lambda \in \Lambda}$ with homomorphisms $\pi_{\lambda,\mu} : G_\lambda \to G_\mu$ whenever $\lambda \geq \mu$ such that $\pi_{\lambda,\lambda} = \mathrm{id}_{G_\lambda}$ and $\pi_{\mu,\nu} \circ \pi_{\lambda,\mu} = \pi_{\lambda,\nu}$ whenever $\lambda \geq \mu \geq \nu$. The **inverse limit**

$$\varprojlim G_\lambda = \varprojlim (G_\lambda)_{\lambda \in \Lambda}$$

is the subgroup of the Cartesian product $\prod_{\lambda \in \Lambda} G_\lambda$ consisting of all elements $(g_\lambda)_{\lambda \in \Lambda}$ with $\pi_{\lambda,\mu}(g_\lambda) = g_\mu$ whenever $\lambda \geq \mu$. If every G_λ is a finite group, then we furnish $\prod_{\lambda \in \Lambda} G_\lambda$ with the product topology of discrete spaces. In this way, $\varprojlim G_\lambda$ with the induced topology becomes a **topological group**, that is, a group which is a topological space such that group multiplication and inversion are continuous. Now a **pro-p group** can be defined as a group which is isomorphic as a topological group to an inverse limit of finite p-groups.

The prototype of a pro-p group is the additive group of the p-adic integers $\mathbb{Z}_p$, which is isomorphic to the inverse limit of the additive groups $(\mathbb{Z}/p^n\mathbb{Z})_{n \in \mathbb{N}}$ where all homomorphisms are projections. Clearly, a finite group is pro-p if and only if it is a p-group.

Every pro-p group G is a $\mathbb{Z}_p$-**(operator) group** and $z \in \mathbb{Z}_p$ acts on $g \in G$ as

$$g^z = \lim_{n \to \infty} g^{z_n} \quad \text{where} \quad z_n = z \bmod p^n.$$

If $g, h \in G$ with $gh = hg$ and $y, z \in \mathbb{Z}_p$, then $g^{y+z} = g^y g^z$, $g^{yz} = (g^y)^z$, and $(gh)^z = g^z h^z$. In particular, if G is abelian, then G is a $\mathbb{Z}_p$-module.

The group G is **finitely generated** (as a topological group), if there exists a finite subset $X \subseteq G$ such that the only open subgroup of G containing $\langle X \rangle$ is G. For example, $(\mathbb{Z}_p, +)$ is generated (topologically) by $\{1\}$, whereas $\langle 1 \rangle = (\mathbb{Z}, +)$ as abstract groups.

Theorem. *If G and H are pro-p groups and G is finitely generated, then every (abstract) group homomorphism $G \to H$ is continuous.*

This implies that every homomorphism from a finitely generated pro-p group G to a pro-p group H is compatible with the action of $\mathbb{Z}_p$. Moreover, the identity $\mathrm{id}_G \colon G \to G$ is continuous and, hence, the topology of G is determined uniquely by the group structure.

We conclude this paragraph with a lemma concerning $\mathbb{Z}_p$-modules. Recall that a $\mathbb{Z}_p$-module is **free of rank** r if it is isomorphic to $\mathbb{Z}_p^r$ for some $r \in \mathbb{N}$.

Lemma. *Every free $\mathbb{Z}_p$-module of finite rank is pro-p, and every finitely generated abelian pro-p group is isomorphic to $\mathbb{Z}_p^r \times N$ for some finite p-group N and $r \in \mathbb{N}$.*

5.3 Homomorphisms from $T \wedge T$

Throughout this section, we write
$$P = C_p(g) \quad \text{and} \quad T = (\mathbb{Z}_p[\theta], +)$$
where g acts on T via multiplication by θ; that is, the group $P \ltimes T$ is isomorphic to the p-adic space group of maximal class. For positive integers n and e let
$$T_n = (\mathfrak{p}^{n-1}, +) \quad \text{and} \quad A_e = T/T_{e+1}.$$
Unless otherwise noted, we assume that $e \leq \mathfrak{e}_n$, see Definition 2.1. It is the aim of this section to determine the structure of the group of P-homomorphisms $T \wedge T \to T$ and $T/T_n \wedge T/T_n \to A_e$, respectively. These groups are of special interest as they are connected to the construction of skeleton groups, cf. Definition 2.2.

5.3.1 The exterior square $T \wedge T$

Let F be the free $\mathbb{Z}_p$-module generated freely by $T \times T$. We write F multiplicatively and define $U \leq F$ as the $\mathbb{Z}_p$-submodule generated by the elements $(a+b,c)(a,c)^{-1}(b,c)^{-1}$, $(a, b+c)(a,b)^{-1}(a,c)^{-1}$, $(a,b)^z (za,b)^{-1}$, and $(a,b)^z (a, zb)^{-1}$ with $a, b, c \in T$ and $z \in \mathbb{Z}_p$. The **tensor product** of T is defined as the $\mathbb{Z}_p$-module
$$T \otimes T = F/U$$
and we write $a \otimes b = (a,b)U$ for $a, b \in T$. It is a P-module via $(a \otimes b)^g = a^g \otimes b^g$.

The **exterior square** of T is
$$T \wedge T = (T \otimes T)/V$$
where V is the $\mathbb{Z}_p$-submodule of $T \otimes T$ generated by $\{a \otimes a \mid a \in T\}$. We write $a \wedge b = (a \otimes b)V$, and, by definition, $a \wedge b = (b \wedge a)^{-1}$ for all $a, b \in T$. The $\mathbb{Z}_p P$-module structure of $T \wedge T$ is defined by
$$(\theta^i \wedge \theta^j)^{\sum_{l=0}^d a_l g^l} = \prod_{l=0}^d (\theta^{i+l} \wedge \theta^{j+l})^{a_l}.$$
In particular, $T \wedge T$ is not a $\mathbb{Z}_p[\theta]$-module as $(a \wedge b)^{1+\theta+\cdots+\theta^d} \neq (a \wedge b)^0 = 0$ in general.

The following theorem is a modification of [29, Proposition 8.3.5].

5.4 Theorem. *The $\mathbb{Z}_p P$-module $T \wedge T$ is the direct product of a free $\mathbb{Z}_p P$-module of rank $d/2 - 1$ generated freely by $\{1 \wedge \theta^2, \ldots, 1 \wedge \theta^{d/2}\}$, and a free $\mathbb{Z}_p$-module generated by the element $z = \prod_{0 \leq i < j < d} \theta^i \wedge \theta^j$.*

Proof. Clearly, $M = \{\theta^i \wedge \theta^j \mid 0 \leq i < j < d\}$ is a free generating set for $T \wedge T$ as a $\mathbb{Z}_p$-module. As a $\mathbb{Z}_p P$-module, $T \wedge T$ is generated by $\{1 \wedge \theta, \ldots, 1 \wedge \theta^{d/2}\}$ since

$$\theta^i \wedge \theta^j = (1 \wedge \theta^{j-i})^{g^i} \quad \text{and} \quad 1 \wedge \theta^j = (1 \wedge \theta^{p-j})^{-g^j}$$

for all $0 \leq i < j < d$. For $2 \leq i \leq d/2$, we define B_i as the $\mathbb{Z}_p P$-submodule of $T \wedge T$ generated by $\{1 \wedge \theta^i\}$, and $B_1 = \{z^c \mid c \in \mathbb{Z}_p\}$. By definition, $z^g = z$ and B_1 has $\mathbb{Z}_p$-rank 1. If $2 \leq i \leq d/2$, then B_i has $\mathbb{Z}_p P$-rank 1 and a $\mathbb{Z}_p$-basis of B_i is given by

$$N_i = \{1 \wedge \theta^i, \ldots, \theta^{d-i} \wedge \theta^d, \ 1 \wedge \theta^{d-i+1}, \ldots, \theta^{i-1} \wedge \theta^d\}.$$

Now let $N = \bigcup_{i=2}^{d/2} N_i$ and consider a $\mathbb{Z}_p$-linear combination $m = \prod_{u \in N} u^{a_u}$ with $m = 1$. If $M' = M \cap N$, then this linear combination can be written as

$$m = \prod_{u \in M'} u^{a_u} \prod_{j=1}^{d-2} (\theta^j \wedge \theta^d)^{a_j}$$

with $a_1, \ldots, a_{d-2} \in \mathbb{Z}_p$. If $\langle M' \rangle_{\mathbb{Z}_p}$ is the $\mathbb{Z}_p$-submodule of $T \wedge T$ generated by M', then

$$\theta^j \wedge \theta^d \equiv (\theta^{j-1} \wedge \theta^j)(\theta^j \wedge \theta^{j+1})^{-1} \mod \langle M' \rangle_{\mathbb{Z}_p}$$

for all $1 \leq j \leq d-2$. This implies that $a_1 = \ldots = a_{d-2} = 0$ and, moreover, $a_u = 0$ for all $u \in N$ since M is a free generating set. Therefore, $B = \langle N \rangle_{\mathbb{Z}_p}$ is freely generated by N as a $\mathbb{Z}_p$-module, and it is freely generated by $\{1 \wedge \theta^2, \ldots, 1 \wedge \theta^{d/2}\}$ as a $\mathbb{Z}_p P$-module. For $2 \leq k \leq d/2$ let $\alpha_k = g + \ldots + g^{p-k}$. A straightforward computation shows that

$$(1 \wedge \theta)^{g-1} = \prod_{k=2}^{d-1} (1 \wedge \theta^k)^{g^{1-k}} = \prod_{k=2}^{d/2} (1 \wedge \theta^k)^{(g-1)\alpha_k}$$

and $\theta^i \wedge \theta^{i+1} \equiv 1 \wedge \theta \mod B$ for all $0 \leq i \leq d-1$; that is, $z \equiv (1 \wedge \theta)^{d-1} \mod B$. This proves that $\{z\} \cup N$ generates $T \wedge T$ as a $\mathbb{Z}_p$-module. Another computation shows that $(1 \wedge \theta)^a \notin B$ for all $a \in \mathbb{Z}_p \setminus \{0\}$; that is, $B_1 \cap B = \{1\}$. This proves the theorem. $\square$

5.3.2 Homomorphisms (I)

As free $\mathbb{Z}_p$-modules of finite rank, the groups $T \wedge T$ and T are pro-p and, therefore, every homomorphism $T \wedge T \to T$ and $T \wedge T \to A_e$ is a $\mathbb{Z}_p$-module homomorphism.

The element $z \in T \wedge T$ of Theorem 5.4 is a fixed point under the action of P and, by Theorem 5.4, every P-homomorphism $T \wedge T \to T$ is uniquely determined by its values on $1 \wedge \theta^2, \ldots, 1 \wedge \theta^{d/2}$. The elements in A_e fixed by $g \in P$ are precisely those in the unique minimal P-invariant subgroup $A_e^P = T_e/T_{e+1}$.

5.3. Homomorphisms from $T \wedge T$

5.5 Definition. Let $3 \leq k \leq d/2 + 1$ and let $\widehat{z} \in A_e^P$ be a generator.

a) The P-homomorphism $f_k \colon T \wedge T \to T$ is defined by

$$f_k(z) = 0 \quad \text{and} \quad f_k(1 \wedge \theta^{j-1}) = \delta_{j,k} \quad (3 \leq j \leq d/2 + 1)$$

where $\delta_{u,v}$ is the Kronecker delta with $\delta_{u,v} = 1$ if $u = v$ and $\delta_{u,v} = 0$ if $u \neq v$.

b) The P-homomorphism $\widehat{f}_k \colon T \wedge T \to A_e$ is defined as $\widehat{f}_k = \pi \circ f_k$ where $\pi \colon T \to A_e$ is the projection, and $\widehat{f}_2 \colon T \wedge T \to A_e$ is defined by

$$\widehat{f}_2(z) = \widehat{z} \quad \text{and} \quad \widehat{f}_2(1 \wedge \theta^j) = 0 \quad (2 \leq j \leq d/2).$$

It follows from Theorem 5.4 that

$$\widehat{f}_2(1 \wedge \theta) = (d-1)^{-1}\widehat{z} \quad \text{and} \quad f_k(1 \wedge \theta) = \theta + \ldots + \theta^{p+1-k}.$$

We regard $\mathrm{Hom}_P(T \wedge T, T)$ as a $\mathbb{Z}_p[\theta]$-module via $(uf)(v) = u(f(v))$ where $u \in \mathbb{Z}_p[\theta]$, $f \in \mathrm{Hom}_P(T \wedge T, T)$, and $v \in T \wedge T$. The following corollary is from [29, Theorem 8.3.7].

5.6 Corollary. a) *As a $\mathbb{Z}_p[\theta]$-module, $\mathrm{Hom}_P(T \wedge T, T)$ is freely generated by $f_3, \ldots, f_{d/2+1}$.*

b) *$\mathrm{Hom}_P(T \wedge T, A_e)$ is the direct sum of $d/2 - 1$ summands isomorphic to A_e, generated by the homomorphisms $\widehat{f}_3, \ldots, \widehat{f}_{d/2+1}$, and a summand of order p generated by $\widehat{f}_2$.*

Proof. By Theorem 5.4, every P-homomorphism $T \wedge T \to T$ is uniquely determined by its values on $1 \wedge \theta^2, \ldots, 1 \wedge \theta^{d/2}$; that is, every element of $\mathrm{Hom}_P(T \wedge T, T)$ can be written uniquely as a $\mathbb{Z}_p[\theta]$-linear combination of $f_3, \ldots, f_{d/2+1}$. Every P-homomorphism from $T \wedge T$ to A_e maps z onto an element of A_e^P. $\square$

The projection $A_{e+1} \to A_e$ has kernel A_{e+1}^P, and Corollary 5.6 proves the following.

5.7 Corollary. *A P-homomorphism $f \colon T \wedge T \to A_e$ can be lifted to a P-homomorphism $T \wedge T \to A_{e+1}$ and $T \wedge T \to T$, respectively, if and only if $f(z) = 0$.*

We now show that $\mathrm{Hom}_P(T/T_n \wedge T/T_n, A_e)$ and $\mathrm{Hom}_P(T \wedge T, A_e)$ can be identified if $n \geq p$ and, as assumed, $e \leq \mathfrak{e}_n$. In particular, a P-homomorphism $T \wedge T \to A_e$ can be applied to elements of $T/T_n \wedge T/T_n$.

5.8 Lemma. *If $n \geq p$ and $e \leq \lfloor \frac{n-1}{d} \rfloor d$, then*

$$\mathrm{Hom}_P(T \wedge T, A_e) \to \mathrm{Hom}_P(T/T_n \wedge T/T_n, A_e), \quad f \mapsto \widetilde{f},$$

with $\widetilde{f} \colon (a + T_n) \wedge (b + T_n) \mapsto f(a \wedge b)$ is an isomorphism.

Proof. It follows from $e \leq \lfloor \frac{n-1}{d} \rfloor d$ that there exists an integer $x \geq 1$ such that $u \in T_{xd+1}$ and $p^x v \in T_{e+1}$ for all $u \in T_n$ and $v \in T$. This shows that $f(a \wedge b) = 0$ whenever $a \in T_n$ or $b \in T_n$, that is, $\widetilde{f}$ is well-defined. Clearly, the mapping $f \to \widetilde{f}$ is an injective homomorphism. It is surjective since every P-homomorphism $T/T_n \wedge T/T_n \to A_e$ is already determined by its values on $\prod_{0 \leq i < j < d}(\theta^i + T_n) \wedge (\theta^j + T_n)$ and $(1 + T_n) \wedge (\theta^j + T_n)$ for $2 \leq j \leq d/2$. $\square$

We end this paragraph with an easy observation. Recall that T is a cyclic $\mathbb{Z}_p[\theta]$-module.

5.9 Lemma. *A P-homomorphism $f \colon T \wedge T \to T$ is surjective if and only if $\mathrm{im}\, f \not\subseteq T_2$.*

5.3.3 Homomorphisms (II)

It is proved in Corollary 5.6 that every P-homomorphism $f\colon T\wedge T\to T$ can be written as a $\mathbb{Z}_p[\theta]$-linear combination

$$f = \sum_{k=3}^{d/2+1} c_k f_k.$$

We now introduce a new set of P-homomorphisms describing $\mathrm{Hom}_P(T\wedge T, T)$, cf. [29, Theorem 8.3.1]. For this purpose, for an integer j with $p\nmid j$ we consider the ring automorphism $\sigma_j\colon \mathbb{Z}_p[\theta]\to\mathbb{Z}_p[\theta]$ defined by

$$\sigma_j\colon \theta\mapsto \theta^{j\bmod p}.$$

Definition. If $2\le a\le d/2$, then the P-homomorphism $F_a\colon T\wedge T\to T$ is defined by

$$F_a\colon T\wedge T\to T,\quad x\wedge y\mapsto \sigma_a(x)\sigma_{1-a}(y) - \sigma_a(y)\sigma_{1-a}(x).$$

By definition, the image of F_a lies in the maximal ideal $\mathfrak{p}$ of T for all $2\le a\le d/2$; that is, the $\mathbb{Z}_p[\theta]$-submodule of $\mathrm{Hom}_P(T\wedge T, T)$ generated by $\{F_2,\ldots,F_{d/2}\}$ is a proper submodule. However, it is shown in the following lemma that every P-homomorphism $T\wedge T\to T$ can be written as a $\mathbb{Q}_p(\theta)$-linear combination of $F_2,\ldots,F_{d/2}$, where $uF_a\colon T\wedge T\to \mathbb{Q}_p(\theta)$ with $u\in\mathbb{Q}_p(\theta)$ is defined as usual.

5.10 Lemma. *Every P-homomorphism $f\colon T\wedge T\to T$ can be written uniquely as*

$$f = \sum_{k=2}^{d/2} c_k F_k$$

with $c_2,\ldots,c_{d/2}\in\mathfrak{p}^{p^\star}$ where $p^\star = -\frac{(p-3)^2}{4}$. If f is surjective, then $c_a\in\mathfrak{p}^{-1}\setminus T$ for some a.

Proof. For all $2\le a\le d/2$, the homomorphism F_a can be written uniquely as

$$F_a = \sum_{k=2}^{d/2}(\theta^{(1-a)k} - \theta^{ka})f_{k+1};$$

that is, the change matrix from $B_1 = \{F_2,\ldots,F_{d/2}\}$ to $B_2 = \{f_3,\ldots,f_{d/2+1}\}$ is

$$M = (\theta^{(1-a)k} - \theta^{ka})_{2\le a,k\le d/2}.$$

In [29, Theorem 8.3.7], a free generating set B_2' of $\mathrm{Hom}_P(T\wedge T, T)$ is given and it follows from the proof of [29, Proposition 8.3.8] that the basis change matrix from B_1 to B_2' has non-zero determinant lying in $\mathfrak{p}^{-p^\star}\setminus\mathfrak{p}^{-p^\star+1}$. The basis change matrix from B_2' to B_2 is invertible over $\mathbb{Z}_p[\theta]$ and, thus, the determinant of M lies in $\mathfrak{p}^{-p^\star}\setminus\mathfrak{p}^{-p^\star+1}$. □

5.4 The action of p-adic units

Again, we write $T = (\mathbb{Z}_p[\theta], +)$ and $T_n = (\mathfrak{p}^{n-1}, +)$ for $n\ge 1$. Let $P = C_p(g)$ where g acts on T via multiplication by θ. In this section, the group of P-homomorphisms from $T\wedge T$ to T_{e+1} is denoted by

$$H_e = \mathrm{Hom}_P(T\wedge T, T_{e+1}),$$

and, using Corollary 5.7, we identify the group of liftable P-homomorphisms $T\wedge T\to A_e$ with H_0/H_e. We now describe an action of the unit group $\mathcal{U}_p = \mathbb{Z}_p[\theta]^\star$ on H_0, which is reconsidered in Chapter 9. The results of this section are based on [28, 29].

5.4. The action of p-adic units

5.4.1 Definition of the action

We identify the ring automorphism $\sigma_j \colon \mathbb{Z}_p[\theta] \to \mathbb{Z}_p[\theta]$ defined by $\theta \mapsto \theta^{j \bmod p}$ with its continuation to a $\mathbb{Q}_p$-linear transformation of $\mathbb{Q}_p(\theta)$. We make use of Lemma 5.10 and let a unit $z \in \mathcal{U}_p$ act on $\mathrm{Hom}_P(T \wedge T, T)$ via

$$(cF_a)^{\tilde{z}} = z^{-1}\sigma_a(z)\sigma_{1-a}(z)cF_a \quad (c \in \mathbb{Q}_p(\theta),\ 2 \leq a \leq d/2).$$

A motivation for this action is given in Chapter 9. With respect to this definition, it is natural to examine the following homomorphism.

Definition. For $2 \leq a \leq d/2$ let

$$\rho_a \colon \mathcal{U}_p \to \mathcal{U}_p, \quad z \mapsto z^{-1}\sigma_a(z)\sigma_{1-a}(z).$$

Recall that the group of units is

$$\mathcal{U}_p = C_{p-1}(\omega) \times C_p(\theta) \times \mathcal{U}_p^{(2)}$$

where $\omega \in \mathbb{Z}_p$ is a primitive $(p-1)$-th root of unity and $\mathcal{U}_p^{(2)} = 1 + \mathfrak{p}^2$, see Section 5.1.3. By definition, $\rho_a(\theta) = 1$ and $\rho_a(\omega) = \omega$ and, thus, it remains to investigate the restriction of ρ_a to the group of second one-units. The logarithm mapping of Theorem 5.2 preserves the action of σ_j and, thus, it can be used to translate ρ_a to an additive mapping τ_a from $(\mathfrak{p}^2, +)$ to $(\mathfrak{p}^2, +)$, which allows us to use linear algebra.

Definition. For $2 \leq a \leq d/2$ let

$$\tau_a \colon \mathfrak{p}^2 \to \mathfrak{p}^2, \quad z \mapsto -z + \sigma_a(z) + \sigma_{1-a}(z).$$

We also identify τ_a with its continuation to a $\mathbb{Q}_p$-linear transformation of $\mathbb{Q}_p(\theta)$. We conclude this paragraph with an important observation on $H_e = \mathrm{Hom}_P(T \wedge T, T_{e+1})$.

Lemma. *The groups $H_0, H_1, \ldots$ are invariant under the action of $\mathcal{U}_p$.*

Proof. If $z \in \mathcal{U}_p$, then $F_a^{\tilde{z}}$ maps $u \wedge v \in T \wedge T$ onto $z^{-1}F_a(zu \wedge zv)$. Thus, if $f \colon T \wedge T \to T_{e+1}$ is a P-homomorphism, then $f^{\tilde{z}} = \mu_z^{-1} \circ f \circ (\mu_z \wedge \mu_z)$ where μ_z is the automorphism of T defined by multiplication by z. This shows that $f^{\tilde{z}} \in H_e$. $\square$

5.4.2 Eigenvalues of τ_a

We show that τ_a has p-adic eigenvalues and can be diagonalized. Let $2 \leq r \leq d$ be a primitive $(p-1)$-th root of unity in $\mathbb{F}_p$ such that $r \equiv \omega \bmod p$, see Section 5.1.1. The results of this paragraph are all from [28].

Lemma. *Let $2 \leq k \leq p$.*
a) *The eigenvalues of σ_r are $\{\omega^0, \omega^1, \ldots, \omega^{d-1}\}$.*
b) *If $v \in \mathfrak{p}^k \setminus \mathfrak{p}^{k+1}$ is an eigenvector of σ_r, then the corresponding eigenvalue is ω^k.*
c) *There is an eigenvector $v_k \in \mathfrak{p}^k \setminus \mathfrak{p}^{k+1}$ of σ_r with eigenvalue ω^k.*
d) *If $a \equiv r^i \bmod p$ with $1 \leq i \leq d$, then v_k is an eigenvector of σ_a with eigenvalue ω^{ik}.*

Proof. a) The minimal polynomial $\mu(X)$ of σ_r divides $X^d - 1$. Since the minimal polynomial of θ over $\mathbb{Q}_p$ has degree d, it follows from $\mu(\sigma_r)(\theta) = 0$ that $\mu(X) = X^d - 1$. The roots of $\mu(X)$ in $\mathbb{Z}_p$ are exactly the $(p-1)$-th roots of unity $1, \omega, \omega^2, \ldots, \omega^{d-1}$.

b) We can assume that $v \equiv (\theta - 1)^k \bmod (\mathfrak{p}^{k+1}, +)$. Then

$$\sigma_r(v) \equiv (\theta - 1)^k (1 + \theta + \ldots + \theta^{r-1})^k \equiv r^k (\theta - 1)^k \equiv \omega^k v \bmod (\mathfrak{p}^{k+1}, +).$$

By part a), the eigenvalue corresponding to v is ω^k.

c) By part a), there is an eigenvector w_k of σ_r with eigenvalue ω^k. Let $w_k \in \mathfrak{p}^l \setminus \mathfrak{p}^{l+1}$ and write $l = xd + i$ with integers $x \geq 0$ and $0 \leq i \leq d-1$; that is, $\mathfrak{p}^l = p^x \mathfrak{p}^i$. Thus, the quotient of w_k by an appropriate power of p is an eigenvector lying in $\mathfrak{p}^k / \mathfrak{p}^{k+1}$.

d) This follows from $\sigma_a = \sigma_r^i$. □

We now show that τ_a can be diagonalized.

Definition. Let $2 \leq k \leq p$.

a) Let $v_k \in \mathfrak{p}^k \setminus \mathfrak{p}^{k+1}$ be an eigenvector of σ_r with eigenvalue ω^k.
b) Let V_k be the intersection of $\mathfrak{p}^2$ with the subspace generated by v_k.
c) Let W_k be the $\mathbb{Z}_p$-submodule of $\mathcal{U}_p^{(2)}$ generated by $\exp(v_k)$.

Corollary. Let $2 \leq a \leq p$ with $a \equiv r^i \bmod p$ and $1 - a \equiv r^j \bmod p$.

a) The element v_k is an eigenvector of τ_a with eigenvalue $\omega^{ik} + \omega^{jk} - 1$ for all $2 \leq k \leq p$.
b) The group $(\mathfrak{p}^2, +)$ is the direct sum of the τ_a-invariant $\mathbb{Z}_p$-modules $V_2, \ldots, V_p$.
c) If $a_2 v_2 + \ldots + a_p v_p \in \mathfrak{p}^e$ with $a_2, \ldots, a_p \in \mathbb{Z}_p$, then $a_k v_k \in \mathfrak{p}^e$ for all $2 \leq k \leq p$.

Proof. a) This follows from the definition.

b) It follows from part a) that $W = \{v_2, \ldots, v_p\}$ is a $\mathbb{Q}_p$-basis of $\mathbb{Q}_p(\theta)$. Thus, the set W is $\mathbb{Z}_p$-linearly independent and its $\mathbb{Z}_p$-span lies in $\mathfrak{p}^2$. Now the assertion follows from $p^x v_k \in \mathfrak{p}^{k+xd} \setminus \mathfrak{p}^{k+xd+1}$ for all integers $x \geq 0$ and $2 \leq k \leq p$.

c) A $\mathbb{Z}_p$-basis of $\mathfrak{p}^e$ is given by $\{p^{x_2} v_2, \ldots, p^{x_p} v_p\}$ for suitable integers $x_2, \ldots, x_p \geq 0$. □

Using the exponential mapping, the results of the previous corollary can be transferred to the group of second one-units.

5.11 Corollary. Let $2 \leq a \leq d/2$ with $a \equiv r^i \bmod p$ and $1 - a \equiv r^j \bmod p$.

a) The group $\mathcal{U}_p^{(2)}$ is the direct product of the ρ_a-invariant $\mathbb{Z}_p$-modules $W_2, \ldots, W_p$.
b) If $\exp(v_2)^{a_2} \ldots \exp(v_p)^{a_p} \in 1 + \mathfrak{p}^e$ for some $a_2, \ldots, a_p \in \mathbb{Z}_p$, then $\exp(v_k)^{a_k} \in 1 + \mathfrak{p}^e$ for all $2 \leq k \leq p$.

The following lemma can be found in [28, Proposition 2.6].

5.12 Lemma. Let $2 \leq a \leq d/2$. If $p \equiv 5 \bmod 6$, then $\ker \tau_a$ is trivial. If $p \equiv 1 \bmod 6$ and a is the unique solution of the congruence $X^2 - X + 1 \equiv 0 \bmod p$ lying in the given range, then $\ker \tau_a$ has dimension $d/3$.

5.4. The action of p-adic units

Proof. Let $a \equiv r^i \bmod p$ and $1 - a \equiv r^j \bmod p$. By definition, the dimension of $\ker \tau_a$ is the number of values of k in the range $0 \leq k \leq d-1$ for which $\omega^{ik} + \omega^{jk} = 1$. For such a value of k, we regard ω as a complex number, see Remark 5.1. Then ω^{ik}, 0, and 1 form the vertices of an equilateral triangle in the complex plane. This happens if and only if ω^{ik} is a primitive 6-th root of unity and $\omega^{jk} = \omega^{-ik}$. If ω^{ik} is a primitive 6-th root of unity, then 6 divides $p - 1$. Thus $\ker \tau_a$ is trivial if $p \equiv 5 \bmod 6$.

Now let $p \equiv 1 \bmod 6$. The primitive 6-th roots of unity modulo p are exactly the solutions of the congruence $X(1 - X) \equiv 1 \bmod p$. If $2 \leq a \leq d$ is such a solution, then $p + 1 - a$ is another one. In particular, $a \neq d/2 - 1$ and, thus, there is exactly one solution in the given range. It follows from $a \equiv \omega^i \bmod p$ that ω^i is a primitive 6-th root of unity modulo p and, thus, $1 - \omega^i \equiv \omega^{-i} \bmod p$. Hence, $1 - a \equiv \omega^{-i} \bmod p$ and τ_a has eigenvalues $\{\omega^{ik} + \omega^{-ik} - 1 \mid 0 \leq k \leq d-1\}$. As above, $\omega^{ik} + \omega^{-ik} = 1$ if and only if ω^{ik} is a primitive 6-th root of unity; that is, if and only if k and 6 are coprime. In $\{0, 1, \ldots, d-1\}$ there are $d/2$ even integers and $d/6$ odd integers divisible by 3. This proves the lemma. $\square$

5.4.3 Stabilizers

Motivated by Lemma 5.12, we consider a prime $p \equiv 5 \bmod 6$ in this paragraph. Recall that the $\mathbb{Q}_p$-linear mapping

$$\tau_a : \mathfrak{p}^2 \to \mathfrak{p}^2, \quad z \mapsto -z + \sigma_a(z) + \sigma_{1-a}(z),$$

has eigenvectors $v_2, \ldots, v_p$ for all $2 \leq a \leq d/2$. We denote the eigenvalue of τ_a corresponding to v_k by

$$\omega_{a,k} = \omega^{ik} + \omega^{jk} - 1$$

where $0 \leq i, j < d$ are defined by $a \equiv r^i \bmod p$ and $1 - a \equiv r^j \bmod p$. By Lemma 5.12, there exists an integer $p_{a,k}$ which is maximal with respect to $\omega_{a,k} \equiv 0 \bmod p^{p_{a,k}}$. For $e \geq 2$ and $2 \leq a \leq d/2$ we define

$$v_{a,k,e} = \max\{\lceil (e-k)/d \rceil - p_{a,k}, 0\}.$$

5.13 Lemma. *Let $2 \leq a \leq d/2$ and $u = \exp(v_2)^{a_2} \ldots \exp(v_p)^{a_p}$ with $a_2, \ldots, a_p \in \mathbb{Z}_p$. Then $\rho_a(u)$ lies in $\mathcal{U}_p^{(e)} = 1 + \mathfrak{p}^e$ if and only if $a_k \equiv 0 \bmod p^{v_{a,k,e}}$ for all $2 \leq k \leq p$.*

Proof. It follows from Theorem 5.2, Corollary 5.11, and Lemma 5.12 that $\rho_a(u) \in 1 + \mathfrak{p}^e$ if and only if $\exp(\tau_a(a_2v_2 + \ldots + a_pv_p)) \in 1 + \mathfrak{p}^e$, if and only if $\exp(v_2)^{a_2\omega_{a,2}} \ldots \exp(v_p)^{a_p\omega_{a,p}}$ lies in $1 + \mathfrak{p}^e$, if and only if $a_k\omega_{a,k}v_k \in \mathfrak{p}^e$ for all $2 \leq k \leq p$, if and only if $a_k \in \mathfrak{p}^{e-k-p_{a,k}d}$ for all $2 \leq k \leq p$. $\square$

By Lemma 5.12, Lemma 5.13 does not hold for primes $p \equiv 1 \bmod 6$, which distinguishes the case $p \equiv 5 \bmod 6$ from $p \equiv 1 \bmod 6$. Based on Lemma 5.13, we now provide the main result of this section and consider the stabilizer $\text{Stab}_{\mathcal{U}_p^{(2)}}(f + H_e)$ of a surjective P-homomorphism $f : T \wedge T \to T$. Recall that $p^\star = -(p-3)^2/4$ and, if A is an abelian group, then $A^{[p]} = \{a^p \mid a \in A\}$.

5.14 Theorem. Let $f\colon T\wedge T \to T$ be a surjective P-homomorphism. There is a positive integer $e_0 = e_0(p)$ such that

$$\operatorname{Stab}_{\mathcal{U}_p^{(2)}}(f + H_e)^{[p]} = \operatorname{Stab}_{\mathcal{U}_p^{(2)}}(f + H_{e+d})$$

and

$$\operatorname{Stab}_{\mathcal{U}_p^{(2)}}(f + H_e) \leq C_{\mathcal{U}_p^{(2)}}(H_i/H_{i+3d})$$

for all $e \geq e_0$ and $i \geq 0$.

Proof. We write $f = \sum_{a=2}^{d/2} c_a F_a$ and, if $u \in \mathbb{Q}_p(\theta)$, then $\nu(u)$ is defined by $u \in \mathfrak{p}^{\nu(u)} \setminus \mathfrak{p}^{\nu(u)+1}$ with $\nu(0) = \infty$. First, we consider $u \in \operatorname{Stab}_{\mathcal{U}_p^{(2)}}(f + H_e)$ and define $t_a \in \mathfrak{p}^2$ by

$$\rho_a(u) = 1 + t_a \quad (2 \leq a \leq d/2).$$

Then $\sum_a c_a t_a \kappa^{-e} F_a \in H_0$, and Lemma 5.10 shows that $c_a t_a \in \mathfrak{p}^{p^\star + e}$ for all a; that is, $\nu(c_a) + \nu(t_a) \geq p^\star + e$. Since f is surjective, there is a' with $\nu(c_{a'}) < 0$; that is, $t_{a'} \in \mathfrak{p}^{p^\star + e + 1}$ and $\rho_{a'}(u) \in \mathcal{U}_p^{(p^\star + e + 1)}$. Thus, by Lemma 5.13, we can assume that e is chosen large enough (depending only on p) such that $\nu(t_a) \geq (p-1)/4$ for all a; that is, $p^\star + d\nu(t_a) \geq d - 1$. This implies that

$$\nu(c_a) + p\nu(t_a) \geq p^\star + e + d\nu(t_a) \geq e + d - 1$$

for all a; that is, $\sum_a c_a t_a^p F_a \in H_{e+d}$. Clearly, u^j stabilizes $f + H_e$ and, by induction, $\sum_a c_a t_a^j F_a \in H_e$ for all $1 \leq j \leq p$. Now $u^p \in \operatorname{Stab}_{\mathcal{U}_p^{(2)}}(f + H_{e+d})$ follows from

$$f^{\tilde{u}^p} - f = \sum_{a=2}^{d/2} c_a((1+t_a)^p - 1)F_a$$
$$= \sum_{j=1}^{d} \frac{1}{p}\binom{p}{j} \sum_{a=2}^{d/2} pc_a t_a^j F_a + \sum_{a=2}^{d/2} c_a t_a^p F_a \in H_{e+d}.$$

Moreover, we can assume that e is chosen such that $\nu(t_a) \geq 3d - p^\star - 1$ for all a; that is, $u \in C_{\mathcal{U}^{(2)}}(H_i/H_{i+3d})$ for all $i \geq 0$.

We now consider $u \in \operatorname{Stab}_{\mathcal{U}_p^{(2)}}(f + H_{e+d})$ and define $t_a \in \mathfrak{p}^2$ by $\rho_a(u) = 1 + t_a$. Note that $\nu(t_a) + \nu(c_a) \geq p^\star + e + d$ for all a, and there is a' with $\nu(t_{a'}) \geq p^\star + e + d + 1$. By Lemma 5.13, we can assume that e is chosen large enough, depending only on p, such that there exists $v \in \mathcal{U}_p^{(2)}$ with $v^p = u$. We define $d_a \in \mathfrak{p}^2$ by $\rho_a(v) = 1 + d_a$ and it follows from $t_a = \sum_{i=1}^{p} \binom{p}{i} d_a^i$ that $\nu(d_a) = \nu(t_a) - d$ for all a. Thus, we can assume that e is chosen such that $\nu(d_a) \geq 3d - p^\star - 1$ for all a; that is, $v \in C_{\mathcal{U}_p^{(2)}}(H_i/H_{i+3d})$ for all i. Hence, if $\sum_{a=2}^{d/2} h_a F_a \in H_i$, then $\sum_{a=2}^{d/2} h_a d_a^j F_a \in H_{i+3dj}$ for all $i, j \geq 0$. If we write

$$\sum_{a=2}^{d/2} c_a t_a F_a = g_1 + \ldots + g_p \quad \text{with} \quad g_i = \binom{p}{i} \sum_{a=2}^{d/2} c_a d_a^i F_a,$$

then $g_1 + \ldots + g_p \in H_{e+d}$ implies that $g_1 \in H_{e+d}$; that is, $\sum_a c_a d_a F_a \in H_e$. This shows that $v \in \operatorname{Stab}_{\mathcal{U}_p^{(2)}}(f + H_e)$, which proves the lemma. □

5.15 Corollary. *The bound in Theorem 5.14 can be chosen as*

$$e_0 = d^2/2 + d(\mu + 2)$$

where $\mu = \max\{p_{a,k} \mid 2 \leq a \leq d/2,\ 2 \leq k \leq p\}$. *If* $p = 5$, *then* $d^2/2 + d(\mu + 2) = 20$.

5.4. The action of p-adic units

Proof. Let $u \in \text{Stab}_{\mathcal{U}_p^{(2)}}(f + H_e)$ and write $u = \exp(v_2)^{a_2} \ldots \exp(v_p)^{a_p}$ and $\rho_a(u) = 1 + t_a$ for $2 \leq a \leq d/2$. It follows as in the proof of Theorem 5.14, that there exists a' with $a_k \equiv 0 \bmod p^{v_{a',k,p^\star+e}+1}$ for all $2 \leq k \leq p$, see Lemma 5.13. We require that

$$v_{a',k,p^\star+e+1} \geq v_{a,k,3d-p^\star-1}$$

for all $2 \leq k \leq p$ and $2 \leq a \leq d/2$, as then $\nu(t_a) \geq 3d - p^\star - 1 \geq (p-1)/4$ for all a. A straightforward computation shows that this holds if $e \geq e_0$ with e_0 as in the corollary.

Now let $u \in \text{Stab}_{\mathcal{U}_p^{(2)}}(f + H_{e+d})$ and write $u = \exp(v_2)^{a_2} \ldots \exp(v_p)^{a_p}$ and $\rho_a(u) = 1 + t_a$ for $2 \leq a \leq d/2$. Again, there is a' with $a_k \equiv 0 \bmod p^{v_{a',k,p^\star+e+d}+1}$ for all $2 \leq k \leq p$. It can be proved as above that, if $e \geq e_0$, then $\nu(t_a) \geq 4d - p^\star - 1$ for all a. We require that there is $v \in \mathcal{U}_p^{(2)}$ with $v^p = u$; that is, $a_k \equiv 0 \bmod p$ for all $2 \leq k \leq p$. This holds if $v_{a',k,p^\star+e+d+1} > 1$ for all $2 \leq k \leq p$ and it follows readily that this is the case if $e \geq e_0$. Hence, all requirements in the proof of Theorem 5.14 are satisfied if $e \geq e_0$. If $p = 5$ and $r = 2$, then $p_{2,2} = p_{2,3} = p_{2,4} = 0$ and $p_{2,5} = 1$; that is, $\mu = 1$. $\square$

5.16 Remark. In order to specify the bound e_0, one has to determine the value of μ and, therefore, the p-adic valuations of the p-adic eigenvalues

$$\{\omega_{a,k} = \omega^{ik} + \omega^{jk} - 1 \mid 2 \leq k \leq p\}$$

of τ_a for all $2 \leq a \leq d/2$. Recall that i and j are defined by $1 - a \equiv r^j \bmod p$ and $a \equiv r^i \bmod p$ and, thus, a necessary and sufficient condition for $\omega_{a,k} \neq 0$ to be a multiple of p is that $(1-a)^k \equiv 1 - a^k \bmod p$. It follows from Section 5.1.1 that

$$\omega^i \equiv r^{ip^{n-1}} \equiv a^{(p^{n-1})} \bmod p^n \quad \text{and} \quad \omega^j \equiv r^{jp^{n-1}} \equiv (1-a)^{(p^{n-1})} \bmod p^n$$

for all $n \geq 1$. Thus p^n divides $\omega^{ik} + \omega^{jk} - 1$ if and only if p^n divides $a^{kp^{n-1}} + (1-a)^{kp^{n-1}} - 1$. Computer experiments lead to the conjecture that $\mu \leq 2$, that is, $e_0 \leq d^2/2 + 4d$.

6 The space group of maximal class

We consider the p-adic space group S of maximal class in more detail and elaborate its connection with finite p-groups of maximal class. As a first step, we recall basic properties of S and its impact on the coclass graph $\mathcal{G}(p)$. Afterwards, we determine standard polycyclic presentations of an integral version of S and its nilpotent quotients. Finally, we compute some automorphism groups and second cohomology groups. The notation set up in this chapter is used throughout the rest of this thesis.

6.1 Basic definitions

Most results of this section are from [29]. If θ is a primitive p-th root of unity over $\mathbb{Q}_p$, then the p**-adic space group of maximal class** is defined as
$$S = P \ltimes T$$
where
$$P = C_p(\theta) \quad \text{acts on} \quad T = (\mathbb{Z}_p[\theta], +)$$
via multiplication by θ. The groups P and T are the **point group** and **translation subgroup** of S, respectively. Recall that $\mathfrak{p}$ is the ideal of $\mathbb{Z}_p[\theta]$ generated by $\kappa = \theta - 1$, and we define
$$T_n = (\mathfrak{p}^{n-1}, +)$$
for all positive integers n. Now let $n \geq 2$. It is straightforward to prove that
$$T_n = \gamma_n(S) \quad \text{and} \quad T_n = [T_{n-1}, P].$$
The quotient S/T_n has maximal class and we write
$$S_n = P \ltimes T/T_n.$$
If $n \geq 4$, then the 2-step centralizer of S_n is $P_1(S_n) = T/T_n$. For an integer $e \geq 0$ we denote by
$$A_e = T/T_{e+1}$$
the S-module T/T_{e+1} of order p^e. Since T acts trivially on A_e, the S-module A_e can also be considered as a P- and S_n-module, respectively. Thus, the groups S_n and P both act uniserially on A_e with series $A_e > T_2/T_{e+1} > \ldots > T_{e+1}/T_{e+1} = \{0\}$.

Recall that $d = p - 1$ is the **dimension of S**, that is, the $\mathbb{Z}_p$-rank of T. We denote by
$$n = \mathfrak{r}d + \mathfrak{i}$$
the decomposition with integers $\mathfrak{r} \geq 0$ and $1 \leq \mathfrak{i} \leq d$. The next lemma follows from the results of Section 5.1.

6.1 Lemma. *Let $n \geq 2$ and $e \geq 1$ be integers and write $n = \mathfrak{r}d + \mathfrak{i}$.*
a) *Multiplication by p is a P-module isomorphism $T_e \to T_{e+d}$, $u \mapsto pu$.*
b) *As a $\mathbb{Z}_p$-module, T_n is generated freely by*

$$\{p^{\mathfrak{r}} - w_{n,1},\ p^{\mathfrak{r}}\theta - w_{n,2},\ \ldots,\ p^{\mathfrak{r}}\theta^{d-\mathfrak{i}} - w_{n,d-\mathfrak{i}+1},\ p^{\mathfrak{r}+1}\theta^{d-\mathfrak{i}+1},\ \ldots,\ p^{\mathfrak{r}+1}\theta^{d-1}\}$$

where $w_{n,1},\ldots,w_{n,d-\mathfrak{i}+1}$ are certain $\mathbb{Z}$-linear combinations of $\{p^{\mathfrak{r}}\theta^{d-\mathfrak{i}+1},\ldots,p^{\mathfrak{r}}\theta^{d-1}\}$.
c) *The exponent of T/T_n is $p^{\mathfrak{r}}$ if $\mathfrak{i} = 1$ and $p^{\mathfrak{r}+1}$ if $2 \leq \mathfrak{i} \leq d$.*
d) *The exponent of A_e with $e \leq \left\lfloor \frac{n-1}{d} \right\rfloor d$ is a divisor of $p^{\mathfrak{r}}$.*

Proof. We prove only part b). Parts c) and d) can be deduced from b), and part a) follows from the definition. Since $T_n = p^{\mathfrak{r}} T_{\mathfrak{i}}$, the $\mathbb{Z}_p$-module T_n is generated freely by

$$\{p^{\mathfrak{r}} - w_{n,1},\ p^{\mathfrak{r}}\theta - w_{n,2},\ \ldots,\ p^{\mathfrak{r}}\theta^{d-\mathfrak{i}} - w_{n,d-\mathfrak{i}+1},\ p^{\mathfrak{r}}\theta^{d-\mathfrak{i}+1}\kappa^{\mathfrak{i}-1},\ \ldots,\ p^{\mathfrak{r}}\theta^{d-1}\kappa^{\mathfrak{i}-1}\}$$

where $w_{n,1},\ldots,w_{n,d-\mathfrak{i}+1}$ are certain $\mathbb{Z}$-linear combinations of $\{p^{\mathfrak{r}}\theta^{d-\mathfrak{i}+1},\ldots,p^{\mathfrak{r}}\theta^{d-1}\}$. If H is the $\mathbb{Z}_p$-module generated by the set of the lemma, then $p^{\mathfrak{r}+1}T \leq T_n$ shows that $H \leq T_n$, and $|T/H| \leq |T/T_n| = p^{n-1}$ implies that $T_n = H$. □

6.2 Definition. Let $n = \mathfrak{r}d + \mathfrak{i}$ and e be positive integers. Every element in T_n can be written uniquely as $p^{\mathfrak{r}}\kappa^{\mathfrak{i}-1}u$ for some $u \in T$, and we define $\iota_{n,e}$ as the P-module isomorphism

$$\iota_{n,e}: T_n/T_{n+e} \to A_e,\quad p^{\mathfrak{r}}\kappa^{\mathfrak{i}-1}u + T_{n+e} \mapsto u + T_{e+1}.$$

6.1.1 The integral version of S

As an uncountable group, the p-adic space group S cannot be described by a polycyclic presentation and, hence, we now define an integral, polycyclic version. The **integral version** $S_{\mathbb{Z}}$ of S is defined as

$$S_{\mathbb{Z}} = P \ltimes T_{\mathbb{Z}} \quad \text{with} \quad T_{\mathbb{Z}} = (\mathbb{Z}[\theta], +),$$

where $P = C_p(\theta)$ acts on $T_{\mathbb{Z}}$ via multiplication. Again, the groups P and $T_{\mathbb{Z}}$ are the **point group** and **translation subgroup** of $S_{\mathbb{Z}}$. For $n \geq 2$ we write

$$T_{\mathbb{Z},n} = \gamma_n(S_{\mathbb{Z}}) \quad \text{and} \quad S_{\mathbb{Z},n} = S_{\mathbb{Z}}/T_{\mathbb{Z},n}.$$

Then $T_{\mathbb{Z},n}$ is the principal ideal of $T_{\mathbb{Z}}$ generated by κ^{n-1} and, as a $\mathbb{Z}$-module, it has a generating set as given in Lemma 6.1b). The composition of the embedding $S_{\mathbb{Z}} \to S$ and the projection $S \to S_n$ is surjective with kernel $T_{\mathbb{Z},n}$ and, thus, we can identify $S_{\mathbb{Z},n} \cong S_n$.

The groups $(S_n)_{n \geq 2}$ form an inverse system where all homomorphisms are projections. Its inverse limit is $P \ltimes D$ where D is the inverse limit of $(T_{\mathbb{Z}}/T_{\mathbb{Z},n})_{n \geq 2}$, that is,

$$D = \{(t_1 + T_{\mathbb{Z},2},\ t_1 + t_2 + T_{\mathbb{Z},3},\ \ldots) \mid \forall i: t_i \in T_{\mathbb{Z},i}\}.$$

Recall that $pT_{\mathbb{Z},i} = T_{\mathbb{Z},i+d}$ for all $i \geq 2$ and, thus, it can be shown that

$$D \to T,\quad (t_1 + T_{\mathbb{Z},2},\ t_1 + t_2 + T_{\mathbb{Z},3},\ \ldots) \mapsto \sum_{i \geq 1} t_i,$$

is a well-defined isomorphism of P-modules, that is, $S \cong \varprojlim S_n$ and S is a pro-p group. Up to isomorphism, the groups $P, S_2, S_3, \ldots$ are *all* finite quotients of S. This does not hold for $S_{\mathbb{Z}}$; for example, if $p = 2$, then $S_{\mathbb{Z}} \cong C_2 \ltimes \mathbb{Z}$ has quotients which are not 2-groups.

Lemma. *If $N \leq T$ is a normal subgroup of S of finite index, then $N = T_j$ for some j.*

Proof. It is sufficient to consider $N < T$. It follows from [7, Lemma 1.18] that S/N is a finite p-group and, hence, $\gamma_l(S) = T_l \leq N$ for some $l \geq 2$. The group N/T_l is a P-invariant subgroup of T/T_l in S_l and, by Lemma 3.4, there is $j \leq l$ with $N/T_l = T_j/T_l$; that is, $N = T_j$. □

In summary, it seems more natural to use the p-adic space group S instead of the integral version $S_{\mathbb{Z}}$. On the other hand, the group $S_{\mathbb{Z}}$ can be described by a polycyclic presentation. We use both versions, S and $S_{\mathbb{Z}}$, according to our requirements.

Remark. It is shown in [7, Section 4.6] that every finitely generated pro-p group is isomorphic to a quotient of a finitely generated free pro-p group. Hence, analogously to polycyclic groups, every finitely generated pro-p group can be described by a finite (pro-p) presentation with certain generators and relations. However, we do not adapt the theory presented in Chapter 4 to finitely presented pro-p groups since, for our concerns, it is sufficient to switch to the polycyclic group $S_{\mathbb{Z}}$ if necessary.

6.2 Connection with the graph $\mathcal{G}(p)$

As a motivation for the investigation of S, we recall its impact on the coclass graph $\mathcal{G}(p)$. First, we provide a preliminary lemma, see [29, Lemma 8.2.1].

6.3 Lemma. *Let G be a maximal class group with positive degree of commutativity and refined central series $G > P_1 > \ldots > P_n = \{1\}$. Let $s \in G \setminus P_1$ and $s_1 \in P_1 \setminus P_2$. If $1 \leq k \leq n-2$, or if $k = n-1$ and $s^p = (ss_1)^p = 1$, and if $u > 0$, let $A = P_u/P_{u+k}$. If A is abelian and has order p^k, then A is isomorphic to A_k as a P-module where θ acts on A by conjugation by sP_{u+k}.*

Proof. By Lemma 3.3, the element s^p is central and so conjugation by s induces an action of θ on A. For $i \geq 2$ we inductively define $s_i = [s_{i-1}, s] \in P_i$ and consider T as a $\mathbb{Z}_p$-module generated by $\{\kappa^i \mid i \geq 0\}$ subject to the relations $p\kappa^j + \binom{p}{2}\kappa^{j+1} + \ldots + \binom{p}{p}\kappa^{j+p-1} = 0$ for all $j \geq 0$. Let the mapping $f: T \to A$ be defined by

$$f: a_0 + a_1\kappa + \ldots + a_r\kappa^r \mapsto s_u^{a_0} s_{u+1}^{a_1} \ldots s_{u+r}^{a_r} P_{u+k}$$

for all $a_0, \ldots, a_r \in \mathbb{Z}_p$ and $r \geq k-1$. If f is well-defined, then f obviously is a surjection with kernel T_{k+1}. If $t \in T$, then $f(\theta t) = f(t + \kappa t) = f(t)^{sP_{u+k}}$ as $s_i^s = s_i s_{i+1}$ for all i and, thus, f is a P-homomorphism. It remains to prove that f is well-defined. For all $i \geq u$, it follows by induction from basic commutator identities that

$$(ss_i)^p = s^p s_i^p s_{i+1}^{\binom{p}{2}} \cdots s_{i+p-1}^{\binom{p}{p}} w$$

for some w lying in the normal subgroup of G generated by the set of all commutators in $\{s, s_i\}$ of weight at least 2 in s_i, see [29, Corollary 1.1.7]. Since P_u/P_{u+k} is abelian, this normal subgroup is trivial and so is w. We now show that $s^{-p}(ss_i)^p \in P_{u+k}$ for all $i \geq u$. If $i > 1$, then $s_i \in P_2$ and $s^p = (ss_i)^p$ by Lemma 3.3. Now let $i = 1$ and $u = 1$.

If $1 \leq k \leq n-2$, then $u+k \leq n-1$ and $s^p, (ss_1)^p \in P_{n-1} \leq P_{u+k}$ by Lemma 3.3. If $k = n-1$, then $s^p = (ss_1)^p = 1$ is given. This shows that

$$s_i^p s_{i+1}^{\binom{p}{2}} \cdots s_{i+p-1}^{\binom{p}{p}} \in P_{u+k}$$

for all $i \geq u$; that is, f is well-defined. □

The proof of the following theorem is from [29, Proposition 8.2.3], and it implies the subsequent corollary.

6.4 Theorem. *The graph $\mathcal{G}(p)$ consists of the cyclic group of order p^2 and an infinite coclass tree $\mathcal{T}$ with root S_2. The tree $\mathcal{T}$ has a unique maximal infinite path $S_2 \to S_3 \to \ldots$*

Proof. Except for the cyclic group of order p^2, every maximal class group is connected in $\mathcal{G}(p)$ to the group S_2. It remains to show that the descendant tree $\mathcal{T}$ of S_2 has exactly one infinite path starting at S_2. Obviously, $S_2 \to S_3 \to \ldots$ is such a path in $\mathcal{T}$. Let G be a group in $\mathcal{T}$ which lies on an infinite path. We assume that $|G| = p^n$ with $n > p+1$ so that G has positive degree of commutativity. Let $G > P_1 > \ldots > P_n = \{1\}$ be the refined central series of G and choose $s \in G \setminus P_1$ and $s_1 \in P_1 \setminus P_2$. Since G has infinitely many descendants, it follows from Theorem 3.2 and Lemma 3.3 that P_1 is abelian and $s^p = (ss_1)^p = 1$ in G. Now we can apply Lemma 6.3 and obtain that $P_1 \cong T/T_n$ as P-modules where $\theta \in P$ acts on P_1 by conjugation by s. This shows that $G \cong P \ltimes T/T_n = S_n$. □

6.5 Corollary. *Let G be of maximal class with refined central series $G > P_1 > \ldots > P_n = \{1\}$ where P_1 is abelian and $n > p+1$.*

a) *If G has a descendant in $\mathcal{G}(p)$, then $G \cong S_n$.*

b) *If there exist $s \in G \setminus P_1$ and $s_1 \in P_1 \setminus P_2$ with $s^p = (ss_1)^p = 1$, then $G \cong S_n$.*

Another important property of the space group S is revealed in the following lemma. Recall that $\mathcal{T}_n$ is the n-th body of the tree $\mathcal{T}$, see Section 2.1.1.

6.6 Lemma. *Every group at depth e in $\mathcal{T}_n$ with $n \geq p+1$ is an extension of A_e by S_n.*

Proof. Let G be a group at depth $e > 0$ in the body $\mathcal{T}_n$ with refined central series $G > P_1 > \ldots > P_{n+e} = \{1\}$. By Theorem 3.2, the degree of commutativity l of G is positive and hence P_n is abelian as $e \leq \mathfrak{e}_n$. Let $s \in G \setminus P_1$ be arbitrary. It follows from Lemma 6.3 that P_n and A_e are isomorphic as P-modules where $\theta \in P$ acts on P_n by conjugation by s. Since $e \leq \mathfrak{e}_n$ and $|G| = p^{n+e}$, it can be deduced from Theorem 3.2 that $l \geq e-1$ and, hence, P_1/P_n acts trivially on P_n. The group S_n acts on P_n via $S_n \cong G/P_n$ and, thus, $P_n \cong A_e$ as S_n-modules. □

This shows that every group at depth e in the body $\mathcal{T}_n$ with $n \geq p+1$ can be described by a certain element of the second cohomology group $H^2(S_n, A_e)$. This is the basis for our further investigations and we determine $H^2(S_n, A_e)$ in a later section. The proof of Lemma 6.6 cannot be extended to groups at depth $e > \mathfrak{e}_n$ in the branch $\mathcal{B}_n$, which is the reason why we restrict attention to the groups in the body $\mathcal{T}_n$.

We conclude this paragraph with a preliminary lemma.

6.7 Lemma. *Let G be an extension of A_e by S_n.*
a) *Let $N \trianglelefteq G$ such that G/N has maximal class. If $N < A_e$ or if $N = A_e = \gamma_m(G)$ for some m, then G has maximal class.*
b) *The group G has maximal class if and only if $\gamma_n(G) = A_e$.*

Proof. Part b) follows from part a) and Lemma 3.1. We now prove a). The group G acts uniserially on A_e with series $A_e = B_1 > \ldots > B_{e+1} = \{0\}$, and N is a term of this series. If $N = A_e$ and $N = \gamma_m(G)$, then G/N has order p^n and $m = n$ as G/N has maximal class. By induction, $B_i = \gamma_{n+i-1}(G)$ and G has maximal class. If $N < A_e$, then $N = B_i$ for some $i > 1$ and $\gamma_n(G/B_i) = B_1/B_i$ as G/B_i has maximal class. This shows that $\gamma_n(G) \leq B_1$ and hence $\gamma_n(G) = B_1$. By induction, $\gamma_{n+j}(G) = B_{1+j}$, and G has maximal class. □

6.3 Standard presentations

It follows from Lemma 6.6 that every group at depth e in the body $\mathcal{T}_n$ is an extension of A_e by S_n, and we want to use the cohomological methods described in Section 4.2.3 to examine these extensions. For this purpose, we now define standard p.c.p.s for $S_{\mathbb{Z}}$ and its quotients S_n for $n \geq p$. These presentations are used throughout the rest of this thesis and we use Fraktur characters to increase their recognition value. First, we fix an abstract generating set
$$\mathcal{S} = \{\mathfrak{g}, \mathfrak{t}_1, \ldots, \mathfrak{t}_d\}$$
for the group $S_{\mathbb{Z}}$ where $\mathfrak{g}$ corresponds to the generator θ of the point group P and $\{\mathfrak{t}_1, \ldots, \mathfrak{t}_d\}$ corresponds to the $\mathbb{Z}$-basis $\{1, \theta, \ldots, \theta^{d-1}\}$ of the translation subgroup $T_{\mathbb{Z}}$. As a set of relations we choose
$$\mathcal{R} = \{\mathfrak{g}^p = 1, \quad \mathfrak{t}_k^{\mathfrak{t}_l} = \mathfrak{t}_k, \quad \mathfrak{t}_j^{\mathfrak{g}} = \mathfrak{t}_{j+1}, \quad \mathfrak{t}_d^{\mathfrak{g}} = \mathfrak{t}_1^{-1} \ldots \mathfrak{t}_d^{-1} \mid l < k \text{ and } j < d\}.$$
It is easy to see that $\langle \mathcal{S} \mid \mathcal{R} \rangle$ is a consistent p.c.p. describing the group $S_{\mathbb{Z}}$, and we call it the **standard p.c.p. of $S_{\mathbb{Z}}$**. If we identify $S_{\mathbb{Z}}$ with the group defined by this presentation, then $S_{\mathbb{Z}}$ consists of normalized words in $\mathcal{S}$, and its point group and translation subgroup are $P = \langle \mathfrak{g} \rangle$ and $T_{\mathbb{Z}} = \langle \mathfrak{t}_1, \ldots, \mathfrak{t}_d \rangle$, respectively. Note that the abelian translation subgroup is now written multiplicatively.

Let $n \geq p$. We write $n = \mathfrak{r}d + \mathfrak{i}$ with integers $\mathfrak{r} \geq 1$ and $1 \leq \mathfrak{i} \leq d$, and $q = p^{\mathfrak{r}}$. By Lemma 6.1, the quotient $S_n = S_{\mathbb{Z}}/T_{\mathbb{Z},n}$ has a consistent p.c.p. on $\mathcal{S}$ with defining relations
$$\mathcal{R}_n = \{\mathfrak{g}^p = 1, \quad \mathfrak{t}_k^{\mathfrak{t}_l} = \mathfrak{t}_k, \quad \mathfrak{t}_j^{\mathfrak{g}} = \mathfrak{t}_{j+1}, \quad \mathfrak{t}_d^{\mathfrak{g}} = \mathfrak{t}_1^{l_{n,1}} \ldots \mathfrak{t}_d^{l_{n,d}} \mid l < k \text{ and } j < d\} \cup$$
$$\{\mathfrak{t}_1^q = w_{n,1}, \ldots, \mathfrak{t}_{d-\mathfrak{i}+1}^q = w_{n,d-\mathfrak{i}+1}, \quad \mathfrak{t}_{d-\mathfrak{i}+2}^{pq} = 1, \ldots, \mathfrak{t}_d^{pq} = 1\}$$
where $w_{n,1}, \ldots, w_{n,d-\mathfrak{i}+1}$ are certain normalized words in $\{\mathfrak{t}_{d-\mathfrak{i}+2}^q, \ldots, \mathfrak{t}_d^q\}$, and $\mathfrak{t}_1^{l_{n,1}} \ldots \mathfrak{t}_d^{l_{n,d}}$ is the normalized word corresponding to the element $\mathfrak{t}_1^{-1} \ldots \mathfrak{t}_d^{-1}$. We call this presentation the **standard p.c.p. of S_n**.

For $e \geq 1$ we fix a consistent p.c.p. $\langle \mathcal{A}_e \mid \mathcal{C}_e \rangle$ of $A_e = T/T_{e+1}$ such that $\mathcal{S}$ and $\mathcal{A}_e$ are disjoint, and we call it the **standard p.c.p. of A_e**. If $e \geq d$, then we make use of the standard p.c.p. of S_{e+1} and define $\mathcal{A}_e = \mathcal{A}$ where $\mathcal{A} = \{\mathfrak{a}_1, \ldots, \mathfrak{a}_d\}$ corresponds to the generating set $\{\mathfrak{t}_1, \ldots, \mathfrak{t}_d\}$ of $T_{\mathbb{Z}}$ such that the relations $\mathcal{C}_e$ correspond to a subset of $\mathcal{R}_{e+1}$.

For $e \geq 1$ let $\mathcal{M}_e$ be the set of words in $\mathcal{S} \cup \mathcal{A}_e$ which consists of the conjugate relations describing the $S_{\mathbb{Z}}$-module structure of A_e, cf. Section 4.2.3.

The next definition summarizes the definitions of S and $S_\mathbb{Z}$ which are used in this thesis. Recall that $\mathfrak{G} \in \mathrm{GL}(d, \mathbb{Z}_p)$ is the companion matrix of $1 + X + \ldots + X^d$, see Section 5.1.4.

6.8 Definition. a) The group $S_\mathbb{Z}$ can be regarded as $S_\mathbb{Z} = C_p(\theta) \ltimes \mathbb{Z}[\theta]$ or $S_\mathbb{Z} = \langle \mathcal{S} \mid \mathcal{R} \rangle$. Unless otherwise noted, we assume that $S_\mathbb{Z} = \langle \mathcal{S} \mid \mathcal{R} \rangle$.

b) We abuse notation and identify $\mathfrak{g}$ and $\{\mathfrak{t}_1, \ldots, \mathfrak{t}_d\}$ with the generator θ of P and the $\mathbb{Z}_p$-basis $\{1, \theta, \ldots, \theta^{d-1}\}$ of T, respectively. Thus, S is isomorphic to the group $\mathcal{S}_{\mathbb{Z}_p}$ with elements $\{\mathfrak{g}^{e_0} \mathfrak{t}_1^{e_1} \ldots \mathfrak{t}_d^{e_d} \mid 0 \leq e_0 < d,\ e_1 \ldots, e_d \in \mathbb{Z}_p\}$ and multiplication defined by S. The group S can be regarded as

$$S = C_p(\theta) \ltimes (\mathbb{Z}_p[\theta], +) \quad \text{or} \quad S = C_p(\mathfrak{G}) \ltimes \mathbb{Z}_p^d \quad \text{or} \quad S = \mathcal{S}_{\mathbb{Z}_p}.$$

In all cases, the point group P and translation subgroup T are defined accordingly. Unless otherwise noted, we assume that $S = \mathcal{S}_{\mathbb{Z}_p}$.

c) If $n \geq p$ and $e \geq d$, then S_n and A_e are defined as $S_n = \langle \mathcal{S} \mid \mathcal{R}_n \rangle$ and $A_e = \langle \mathcal{A} \mid \mathcal{C}_e \rangle$.

If $n \geq p$, then every word in $\mathcal{S}$ can be considered as an element of S, $S_\mathbb{Z}$, and S_n, respectively, and we have to make sure that there is no risk of confusion, cf. Remark 4.1. For example, the word $\mathfrak{g}^{-1} \mathfrak{t}_d \mathfrak{g}$ can be considered as

$$\mathfrak{g}^{-1} \mathfrak{t}_d \mathfrak{g} = \mathfrak{t}_1^{-1} \ldots \mathfrak{t}_d^{-1} \text{ in } S \text{ and in } S_\mathbb{Z}, \quad \text{and} \quad \mathfrak{g}^{-1} \mathfrak{t}_d \mathfrak{g} = \mathfrak{t}_1^{l_{n,1}} \ldots \mathfrak{t}_d^{l_{n,d}} \text{ in } S_n.$$

An advantage of this description is the existence of **canonical transversals** $S_n \to S$ and $S_n \to S_\mathbb{Z}$ which map a normalized word $s \in S_n$ onto $s \in S$ and $s \in S_\mathbb{Z}$, respectively.

Recall that the $\mathbb{Z}_p P$-module structure on T now is given by $t^{\sum_{i=0}^{d} a_i \mathfrak{g}^i} = \prod_{i=0}^{d} (t^{\mathfrak{g}^i})^{a_i}$.

6.3.1 The groups $\{S_{n+kd} \mid k \geq 0\}$

We show that the groups in $\{S_{n+kd} \mid k \geq 0\}$ with $n \geq p$ can be described by a single **parameterized presentation** with one integer parameter, that is, by a group presentation whose defining relations have exponents which are arithmetic expressions containing an indeterminate integer as parameter. First, we need some more notation.

If w is a word in a finite alphabet $\{a_1, \ldots, a_m\}$, then we write $w = w(a_1, \ldots, a_m)$ to emphasize the alphabet. If $\{b_1, \ldots, b_m\}$ is some other alphabet, then $w(b_1, \ldots, b_m)$ arises from w by replacing every occurrence of a_i in w by b_i for all $1 \leq i \leq m$.

6.9 Remark. We consider the defining relations in $\mathcal{R}_n$. Then $w_{n,j}$ with $1 \leq j \leq d - \mathfrak{i} + 1$ can be written as

$$w_{n,j} = w_{n,j}(\mathfrak{t}_{d-\mathfrak{i}+2}^{p^\mathfrak{r}}, \ldots, \mathfrak{t}_d^{p^\mathfrak{r}})$$

and, for $k \geq 0$, the word $w_{n+kd,j}$ is given by $w_{n+kd,j} = w_{n,j}(\mathfrak{t}_{d-\mathfrak{i}+2}^{p^{\mathfrak{r}+k}}, \ldots, \mathfrak{t}_d^{p^{\mathfrak{r}+k}})$.

By definition, the exponents $l_{n,1}, \ldots, l_{n,d}$ are all congruent to -1 modulo $p^\mathfrak{r}$ and, by induction, for all $1 \leq j \leq d$ the value of $l_{n+kd,j}$ is

$$l_{n+kd,j} = p^k(l_{n,j} + 1) - 1.$$

A proof is straightforward, but technical, and for details we refer to Section A.1.1. Using these equalities, one can parameterize the relations in $\mathcal{R}_n$ with the integer $k \geq 0$ to describe the relations in $\mathcal{R}_{n+kd}$. This shows that the groups in $\{S_{n+kd} \mid k \geq 0\}$ can be described by a single parameterized presentation with parameter k.

A consequence of Remark 6.9 is the following.

6.10 Remark. If $e \geq d$, then the groups in $\{A_{e+kd} \mid k \geq 0\}$ can be described by a single parameterized presentation with parameter k. Analogously, the sets in $\{\mathcal{M}_{e+kd} \mid k \geq 0\}$, which describe the $S_\mathbb{Z}$-module structures on A_{e+kd} for $k \geq 0$, can be described by a parameterization with parameter k.

6.11 Remark. Let $e \geq d$ and denote by $\pi_k \colon T_\mathbb{Z} \to A_{e+kd}$ the projection. If $u \in T_\mathbb{Z}$, then there exists $k_0 \in \mathbb{N}$ such that the elements in $\{\pi_k(u) \mid k \geq k_0\}$ can be described by a parameterized word with parameter k.

Example. We consider $p = e = 5$ and $n = 6$. Let $k \geq 0$ and $q = 5^k$.

a) The standard p.c.p. of S_6 has generators $\mathcal{S} = \{\mathfrak{g}, \mathfrak{t}_1, \ldots, \mathfrak{t}_4\}$ and relations

$$\mathcal{R}_6 = \{\mathfrak{t}_j^{\mathfrak{t}_i} = \mathfrak{t}_j,\ \mathfrak{t}_l^{\mathfrak{g}} = \mathfrak{t}_{l+1},\ \mathfrak{t}_4^{\mathfrak{g}} = \mathfrak{t}_1^4 \mathfrak{t}_2^4 \mathfrak{t}_3^4 \mathfrak{t}_4^9 \mid i < j \text{ and } l < 4\} \cup$$
$$\{\mathfrak{g}^5 = 1,\ \mathfrak{t}_1^5 = \mathfrak{t}_4^5,\ \mathfrak{t}_2^5 = \mathfrak{t}_4^5,\ \mathfrak{t}_3^5 = \mathfrak{t}_4^5,\ \mathfrak{t}_4^{25} = 1\}.$$

The standard p.c.p. of S_{6+4k} has generating set $\mathcal{S}$ and relations

$$\mathcal{R}_{6+4k} = \{\mathfrak{t}_j^{\mathfrak{t}_i} = \mathfrak{t}_j,\ \mathfrak{t}_l^{\mathfrak{g}} = \mathfrak{t}_{l+1},\ \mathfrak{t}_4^{\mathfrak{g}} = \mathfrak{t}_1^{5q-1} \mathfrak{t}_2^{5q-1} \mathfrak{t}_3^{5q-1} \mathfrak{t}_4^{10q-1} \mid i < j \text{ and } l < 4\} \cup$$
$$\{\mathfrak{g}^5 = 1,\ \mathfrak{t}_1^{5q} = \mathfrak{t}_4^{5q},\ \mathfrak{t}_2^{5q} = \mathfrak{t}_4^{5q},\ \mathfrak{t}_3^{5q} = \mathfrak{t}_4^{5q},\ \mathfrak{t}_4^{25q} = 1\}.$$

b) The standard p.c.p. of A_{5+4k} has generators $\mathcal{A} = \{\mathfrak{a}_1, \ldots, \mathfrak{a}_4\}$ and relations

$$\mathcal{C}_{5+4k} = \{\mathfrak{a}_j^{\mathfrak{a}_i} = \mathfrak{a}_j,\ \mathfrak{a}_1^{5q} = \mathfrak{a}_4^{5q},\ \mathfrak{a}_2^{5q} = \mathfrak{a}_4^{5q},\ \mathfrak{a}_3^{5q} = \mathfrak{a}_4^{5q},\ \mathfrak{a}_4^{25q} = 1 \mid i < j\}.$$

The set $\mathcal{M}_{5+4k}$ is given by

$$\{\mathfrak{a}_i^{\mathfrak{t}_j} = \mathfrak{a}_i,\ \mathfrak{a}_l^{\mathfrak{g}} = \mathfrak{a}_{l+1},\ \mathfrak{a}_4^{\mathfrak{g}} = \mathfrak{a}_1^{5q-1} \mathfrak{a}_2^{5q-1} \mathfrak{a}_3^{5q-1} \mathfrak{a}_4^{10q-1} \mid l < 4 \text{ and } 1 \leq i, j \leq 4\}.$$

c) Let $\pi_k \colon T_\mathbb{Z} \to A_{5+4k}$ be the projection. If $u = \mathfrak{t}_1^{e_1} \mathfrak{t}_2^{-e_2} \mathfrak{t}_3^{e_3} \mathfrak{t}_4^{-e_4} \in T_\mathbb{Z}$ with $e_1, \ldots, e_4 > 0$, then

$$\pi_k(u) = \mathfrak{t}_1^{e_1} \mathfrak{t}_2^{5q-e_2} \mathfrak{t}_3^{e_3} \mathfrak{t}_4^{20q-e_4}$$

for all $k \geq \min\{i \mid 5^{i+1} \geq e_1, e_2, e_3, e_4/4\}$.

6.4 Automorphism groups

The automorphism group of S plays an important role in further investigations and therefore is examined in this section. For this purpose, we consider S as the split extension of its translation subgroup T by its point group P and use the cohomological methods described in Chapter 4.

6.4.1 The automorphism group of S

Recall that the ring automorphism $\sigma_j \colon \mathbb{Z}_p[\theta] \to \mathbb{Z}_p[\theta]$ with $p \nmid j$ is defined by $\theta \mapsto \theta^{j \bmod p}$, and every unit $u \in \mathcal{U}_p$ acts on $\mathbb{Z}_p[\theta]$ via multiplication. We transfer these actions to elements of $\mathrm{Aut}(T)$ for all three descriptions of S, see Definition 6.8.

The group of 1-cocycles $Z^1(P,T)$ consists of the mappings $\delta_t : P \to T$ with $t \in T$ where

$$\delta_t : P \to T, \quad \mathfrak{g}^i \mapsto t^{\frac{\theta^i-1}{\theta-1}}.$$

Note that $\frac{\theta^i-1}{\theta-1}$ is a unit if $p \nmid i$, see Section A.1.2.

6.12 Theorem. *The automorphisms of S are $\{\alpha(j,c,t) \mid 1 \leq j \leq d, \, c \in \mathcal{U}_p, \, t \in T\}$ where*

$$\alpha(j,c,t) \colon S \to S, \quad \mathfrak{g}^l a \mapsto \mathfrak{g}^{lj} \sigma_j(a)^c \delta_t(\mathfrak{g}^{lj}) \quad (0 \leq l \leq d, \, a \in T).$$

Proof. For the proof, we regard $S = P \ltimes T$ with $P = C_p(\mathfrak{G})$ and $T = \mathbb{Z}_p^d$ as in Definition 6.8. The group T is characteristic in S and we consider the homomorphism

$$\phi \colon \mathrm{Aut}(S) \to \mathrm{Aut}(P) \times \mathrm{Aut}(T), \quad \alpha \mapsto (\alpha|_P, \alpha|_T),$$

with $\mathrm{Aut}(P) \cong C_d$ and $\mathrm{Aut}(T) = \mathrm{GL}(d, \mathbb{Z}_p)$. For $t \in T$ let the automorphism $\alpha_t \colon S \to S$ be defined by $\alpha_t|_T = \mathrm{id}_T$ and $\alpha_t \colon (\mathfrak{G}^i, s) \mapsto (\mathfrak{G}^i, s + t(\mathfrak{G}^0 + \ldots + \mathfrak{G}^{i-1}))$ if $1 \leq i \leq d$. It follows from Lemma 4.4 that

$$\ker \phi = \{\alpha_t \colon S \to S \mid t \in T\} \quad \text{and} \quad \mathrm{im}\, \phi = \mathrm{Comp}(P,T),$$

and a preimage of $(\alpha, N) \in \mathrm{Comp}(P,T)$ under ϕ is given by $S \to S$, $(\mathfrak{G}^i, s) \mapsto (\alpha(\mathfrak{G})^i, sN)$. By definition, $(\alpha, N) \in \mathrm{Comp}(P,T)$ if and only if $N \in N_{\mathrm{GL}(d,\mathbb{Z}_p)}(\langle \mathfrak{G} \rangle)$ and $\alpha \in \mathrm{Aut}(P)$ such that $\alpha(\mathfrak{G}) = N^{-1} \mathfrak{G} N$. Now the theorem follows from Lemma 5.3 and a translation to the group S defined as $S = S_{\mathbb{Z}_p}$, see Definition 6.8. □

Remark. a) It follows from the definition that

$$\begin{aligned}
\langle \alpha(j,1,1) \mid 1 \leq j \leq d \rangle &\cong \mathcal{G}(\mathbb{Q}_p(\theta)/\mathbb{Q}_p), \\
\langle \alpha(1,u,1) \mid u \in \mathcal{U}_p \rangle &\cong \mathcal{U}_p, \text{ and} \\
\langle \alpha(1,1,t) \mid t \in T \rangle &\cong Z^1(P,T).
\end{aligned}$$

Using the group actions defined implicitly in Theorem 6.12, the group $\mathrm{Aut}(S)$ is of the isomorphism type $\mathcal{G}(\mathbb{Q}_p(\theta)/\mathbb{Q}_p) \ltimes [\mathcal{U}_p \ltimes Z^1(P,T)]$.

b) The inverse of $\alpha(j,c,t)$ is $\alpha(l, \sigma_l(c)^{-1}, \sigma_l(\delta_{t^{-1}}(g^j))^{\sigma_l(c)^{-1}})$ with $l = j^{-1} \bmod p$.

Corollary. *The inner automorphisms of S are $\{\alpha(1,c,t) \mid c \in C_p(\theta), \, t \in T_2\}$.*

Proof. Let $\mathfrak{g}^j t, \mathfrak{g}^i s \in S$ be arbitrary with $0 \leq i, j \leq d$ and $s, t \in T$. If $i > 0$, then

$$(\mathfrak{g}^i s)^{\mathfrak{g}^j t} = \mathfrak{g}^i s^{\mathfrak{g}^j} (t^{-1})^{\mathfrak{g}^i - 1} = \mathfrak{g}^i s^{\mathfrak{g}^j} (t^{1-\mathfrak{g}})^{1+\mathfrak{g}+\ldots+\mathfrak{g}^{i-1}},$$

and conjugation with $\mathfrak{g}^j t$ induces the automorphism $\alpha(1, \theta^j, t^{1-\mathfrak{g}})$ with $t^{1-\mathfrak{g}} \in T_2$. □

6.4.2 The automorphism group of S_n

6.13 Theorem. *If $n \geq \max\{4,p\}$, then $\mathrm{Aut}(S) \to \mathrm{Aut}(S_n)$, $\alpha \mapsto \alpha|_{S_n}$, is surjective.*

Proof. Let $\alpha \in \mathrm{Aut}(S_n)$ and let $\tau \colon S_n \to S$ be the transversal which maps a normalized word $s \in S_n$ onto $s \in S$. The 2-step centralizer $P_1(S_n) = T/T_n$ is characteristic in S_n and, thus, $\alpha(\mathfrak{g}) = \mathfrak{g}^j a$ for some $1 \leq j \leq d$ and $a \in T/T_n$. The automorphism

$$\alpha(j, 1, \tau(a)^{(1+\theta+\ldots+\theta^{j-1})^{-1}})|_{S_n} \circ \alpha^{-1}$$

fixes $\mathfrak{g} \in S_n$, and we can assume that $\alpha(\mathfrak{g}) = \mathfrak{g}$ and $\alpha(\mathfrak{t}_1) = c$ for some $c \in T/T_n$. Note that there exists $e \in \mathbb{Z}_p[\theta]$ such that c can be written as $c = \mathfrak{t}_1^e$. Since $\mathfrak{t}_l = \mathfrak{t}_1^{\mathfrak{g}^{l-1}}$ for $1 \leq l \leq d$, this implies that $\alpha|_{T/T_n}$ is the multiplication with eT_n, and so $e \notin T_2$. Hence, $e \in \mathcal{U}_p$ and $\alpha(\mathfrak{g}^l t) = \mathfrak{g}^l t^e$ proves that $\alpha = \alpha(1, e, 1)|_{S_n}$. □

We remark that Theorem 6.13 holds for all $n \geq 4$. However, we restrict attention to $n \geq p$ as our proof uses the standard p.c.p. of S_n which we have only defined for $n \geq p$.

Corollary. *If $n \geq \max\{4,p\}$, then the following hold.*
a) $\alpha(j,c,t)|_{S_n} = \mathrm{id}_{S_n}$ *if and only if $j = 1$, $c \in \mathcal{U}_p^{(n-1)}$, and $t \in T_n$.*
b) $\mathrm{Aut}(S_n) = \{\alpha|_{S_n} \colon S_n \to S_n \mid \alpha \in \mathrm{Aut}(S)\}$ *and* $|\mathrm{Aut}(S_n)| = (p-1)^2 p^{2n-3}$.

6.5 Cohomology

Using the methods developed in Section 4.2.3, we now compute the cohomology groups $H^2(S_\mathbb{Z}, A_e)$ and $H^2(S_n, A_e)$ for $n \geq p+1$ and $1 \leq e \leq \mathfrak{e}_n$. First, we recall the relevant definitions and introduce some more notation.

Recall that $S_\mathbb{Z}$ and S_n are the groups defined by their standard p.c.p.s $\langle \mathcal{S} \mid \mathcal{R} \rangle$ and $\langle \mathcal{S} \mid \mathcal{R}_n \rangle$, respectively, with generating set $\mathcal{S} = \{\mathfrak{g}, \mathfrak{t}_1, \ldots, \mathfrak{t}_d\}$. Since we restrict attention to these presentations, we adapt the notation of a tail vector of S_n in A_e. By definition, a tail vector x of S_n in A_e is a list $x = (x_r)_{r \in \mathcal{R}_n}$ of elements in A_e. Now, if $r \in \mathcal{R}_n$ is the conjugate relation of $\mathfrak{t}_j^\mathfrak{g}$ or $\mathfrak{t}_j^{\mathfrak{t}_i}$, then we write $x_{0,j} = x_r$ and $x_{i,j} = x_r$, respectively. The tails of the power relations of $\mathfrak{g}, \mathfrak{t}_1, \ldots, \mathfrak{t}_d$ are denoted by $x_{0,0}, \ldots, x_{d,d}$. Hence, from now on we write $(x_{i,j}) = (x_{i,j})_{0 \leq i \leq j \leq d}$ for $(x_r)_{r \in \mathcal{R}_n}$ and, in a similar way, we treat a tail vector of S in A_e.

Recall that $A_e = \langle \mathcal{A}_e \mid \mathcal{C}_e \rangle$ and the set $\mathcal{M}_e$ contains the conjugate relations describing the S_n-action on A_e. The following definition describes the extension defined by a tail vector of S_n in A_e, cf. Sections 4.2.3 and 6.3.

6.14 Definition. Let $x = (x_{i,j})_{0 \leq i \leq j \leq d}$ be a list of elements in A_e. Then $\mathcal{E}(x)$ is the polycyclic presentation with generating set $\mathcal{S} \cup \mathcal{A}_e$ and relations

$$\{\mathfrak{t}_k^{\mathfrak{t}_l} = \mathfrak{t}_k x_{l,k},\ \mathfrak{t}_j^\mathfrak{g} = \mathfrak{t}_{j+1} x_{0,j},\ \mathfrak{t}_d^\mathfrak{g} = \mathfrak{t}_1^{l_{n,1}} \ldots \mathfrak{t}_d^{l_{n,d}} x_{0,d} \mid l < k \text{ and } j < d\} \cup$$

$$\{\mathfrak{g}^p = x_{0,0},\ \mathfrak{t}_j^{p^s} = w_{n,j} x_{j,j},\ \mathfrak{t}_l^{p^{s+1}} = x_{l,l} \mid j \leq d-\mathfrak{i}+1 < l\} \cup \mathcal{C}_e \cup \mathcal{M}_e.$$

The group defined by $\mathcal{E}(x)$ is denoted by $E(x)$.

6.15 Remark. By the construction in Section 4.2.3, the tail vector $x_\gamma = (x_{i,j})$ defined by the 2-cocycle $\gamma \in Z^2(S_n, A_e)$ consists of the tails

$$x_{0,0} = \prod_{i=1}^{d} \gamma(\mathfrak{g}, \mathfrak{g}^i) = \prod_{i=1}^{d} \gamma(\mathfrak{g}^i, \mathfrak{g})^{\mathfrak{g}^{d-i}},$$
$$x_{i,i} = \prod_{k=1}^{q-1} \gamma(\mathfrak{t}_i^k, \mathfrak{t}_i) \quad \text{for } 1 \leq i \leq d \text{ with } q = \mathrm{relord}(\mathfrak{t}_i),$$
$$x_{i,j} = \gamma(\mathfrak{t}_j, \mathfrak{t}_i)\gamma(\mathfrak{t}_i, \mathfrak{t}_j)^{-1} \quad \text{for } 0 < i < j < p, \text{ and}$$
$$x_{0,i} = \gamma(\mathfrak{t}_i, \mathfrak{g})\gamma(\mathfrak{g}, \mathfrak{t}_i^{\mathfrak{g}})^{-1} \quad \text{for } 0 < i < p.$$

For example, $x_{i,j}$ with $0 < i < j < p$ is defined by $(\mathfrak{t}_i, 1)(\mathfrak{t}_j, x_{i,j}) = (\mathfrak{t}_j, 1)(\mathfrak{t}_i, 1)$ in $E(\gamma)$.

6.16 Remark. Let $\gamma \in Z^2(S_n, T_n/T_{n+e})$ be a 2-cocycle defining an extension $E(\gamma)$ of T_n/T_{n+e} by S_n. We use the P-module isomorphism $\iota_{n,e} \colon T_n/T_{n+e} \to A_e$ of Definition 6.2 to obtain a 2-cocycle $\gamma' = \iota_{n,e} \circ \gamma$ of S_n with coefficients in A_e. The extensions $E(\gamma)$ and $E(\gamma')$ are isomorphic via $(s, a) \mapsto (s, \iota_{n,e}(a))$, and $E(\gamma') \cong E(x_{\gamma'})$ by Lemma 4.6. Thus, every extension of T_n/T_{n+e} by S_n can be described by a tail vector of S_n in A_e.

6.5.1 The second cohomology of $S_\mathbb{Z}$

For an $S_\mathbb{Z}$-module A with $T_\mathbb{Z}$ acting trivially on A we write $A^{S_\mathbb{Z}} = A^P$ for the subgroup of fixed points under this action.

6.17 Lemma. *If A is an $S_\mathbb{Z}$-module with $T_\mathbb{Z}$ acting trivially on A, then*

$$H^2(S_\mathbb{Z}, A) \cong (A^P)^3 \times A^{d/2-1}.$$

Proof. Using the methods described in Section 4.2.3, that is, standard presentations and consistency checks, it is straightforward, but technical, to prove that

$$\mathcal{Z}(S_\mathbb{Z}, A) \cong (A^P)^3 \times A^{d/2-1} \times \mathcal{B}(S_\mathbb{Z}, A) \quad \text{with} \quad \mathcal{B}(S_\mathbb{Z}, A) \cong A^{d-1}.$$

A detailed proof is given in Section A.1.3. □

Remark. An analysis of the proof of Lemma 6.17 shows that the tail vectors $(x_{i,j})$ lying in $\mathcal{Z}(S_\mathbb{Z}, A)$ can be parameterized as follows. The elements

- $x_{1,3}, \ldots, x_{1,d/2+1} \in A$,
- $x_{0,1}, \ldots, x_{0,d-1} \in A$,
- and $k_1, k_2, k_3 \in A^P$

can be chosen arbitrarily and determine

- $x_{0,0} = k_1$,
- $x_{1,2} = k_2 \prod_{k=3}^{d/2+1} x_{1,k}^{\alpha_k}$ with $\alpha_k = \mathfrak{g} + \ldots + \mathfrak{g}^{p-k}$,
- $x_{1,i} = x_{1,d-i+3}^{-\mathfrak{g}^{i-1}}$ for $d/2 + 2 \leq i \leq d$,
- $x_{i,j} = x_{1,j-i+1}^{\mathfrak{g}^{i-1}}$ for $2 \leq i < j \leq d$, and
- $x_{0,d} = k_3 \prod_{k=1}^{d-1} x_{0,k}^{-\beta_k}$ with $\beta_k = 1 + \mathfrak{g} + \ldots + \mathfrak{g}^{d-k}$.

A complement to $\mathcal{B}(S_\mathbb{Z}, A)$ in $\mathcal{Z}(S_\mathbb{Z}, A)$ is generated by the tail vectors with trivial tails $x_{0,1}, \ldots, x_{0,d-1}$.

6.5.2 Bounding the order of $H^2(S_n, A_e)$

Recall that $n \geq p+1$ and $1 \leq e \leq \mathfrak{e}_n$. The point group P of $S_\mathbb{Z}$ acts on a 1-cocycle $\gamma \in Z^1(T_{\mathbb{Z},n}, A_e)$ via

$$\gamma^\mathfrak{g} \colon T_{\mathbb{Z},n} \to A_e, \quad t \mapsto \gamma(t^\mathfrak{g})^{\mathfrak{g}^{-1}},$$

and this action gives rise to an action on the first cohomology group $H^1(T_{\mathbb{Z},n}, A_e)$, cf. Section 4.2.1. The group of fixed points under this action is denoted by $H^1(T_{\mathbb{Z},n}, A_e)^P$.

We start with a preliminary lemma.

6.18 Lemma. $H^1(S_\mathbb{Z}, A_e) \cong H^1(S_n, A_e)$ and $H^1(T_{\mathbb{Z},n}, A_e)^P = \mathrm{Hom}_P(T_{\mathbb{Z},n}, A_e) \cong A_e$.

Proof. We write $A = A_e$. If $\gamma \in Z^1(S_\mathbb{Z}, A)$, then $\gamma(\mathfrak{g}^i t) = \gamma(\mathfrak{g})^{\sum_{k=0}^{i-1} \mathfrak{g}^k} \gamma(t)$ for all $1 \leq i \leq d$ and $t \in T_\mathbb{Z}$, and the restriction $\gamma|_{T_\mathbb{Z}}$ is a homomorphism. In particular, it follows from $\gamma(\mathfrak{g}^{i+j} s^{\mathfrak{g}^j} t) = \gamma(\mathfrak{g}^i s)^{\mathfrak{g}^j} \gamma(\mathfrak{g}^j t)$ for $s, t \in T_\mathbb{Z}$ that $\gamma|_{T_\mathbb{Z}}$ is a P-homomorphism; that is, the values $\gamma(\mathfrak{g})$ and $\gamma(t_1)$ describe γ completely. Conversely, if $a, b \in A$ and $f \in \mathrm{Hom}_P(T_\mathbb{Z}, A)$ is defined by $f(t_1) = b$, then a 1-cocycle $\gamma_{a,b} \in Z^1(S_\mathbb{Z}, A)$ is defined by

$$\gamma_{a,b} \colon S_\mathbb{Z} \to A, \quad \mathfrak{g}^i t \mapsto a^{\sum_{k=0}^{i-1} \mathfrak{g}^k} f(t) \quad (0 \leq i \leq d,\ t \in T_\mathbb{Z}).$$

The mapping $A \times A \to Z^1(S_\mathbb{Z}, A)$, $(a, b) \mapsto \gamma_{a,b}$, is an isomorphism.

Every $a \in A$ defines a P-homomorphism $f_a \colon T_\mathbb{Z} \to A$ via $t_1 \mapsto a$. If $s \in T_{\mathbb{Z},n}$ with $s = t^{(\mathfrak{g}-1)^{n-1}}$ for some $t \in T_\mathbb{Z}$, then $T_{\mathbb{Z},n} \leq T_{\mathbb{Z},e+1}$ implies that $f_a(s) = f_a(t)^{(\mathfrak{g}-1)^{n-1}} = 1$, and f_a induces a P-homomorphism $T_\mathbb{Z}/T_{\mathbb{Z},n} \to A$; that is, $Z^1(S_{\mathbb{Z},n}, T_\mathbb{Z}) \cong A \times A$ follows as above. Now it is easy to deduce that $H^1(S_\mathbb{Z}, A_e) \cong H^1(S_{\mathbb{Z},n}, A_e)$. Note that $H^1(T_{\mathbb{Z},n}, A) = \mathrm{Hom}(T_{\mathbb{Z},n}, A)$ and $H^1(T_{\mathbb{Z},n}, A)^P = \mathrm{Hom}_P(T_{\mathbb{Z},n}, A) \cong A$ as shown above. □

A sequence $M_1 \xrightarrow{\alpha_1} M_2 \xrightarrow{\alpha_2} \ldots$ of groups $M_1, M_2, \ldots$ and group homomorphisms $\alpha_1, \alpha_2, \ldots$ is **exact**, if $\ker \alpha_{j+1} = \mathrm{im}\, \alpha_j$ for all $j \geq 1$. We now use an exact sequence of cohomology groups to prove an upper bound for the order of $H^2(S_n, A_e)$. For this purpose, we introduce some more notation, cf. [66, Section 11.8].

6.19 Definition. a) Let $\tau \colon S_n \to S_\mathbb{Z}$ be the transversal which maps a normalized word $s \in S_n$ onto $\tau(s) = s$, and define $\gamma \colon S_n \times S_n \to T_{\mathbb{Z},n}$ as $(u, v) \mapsto \tau(uv)^{-1} \tau(u) \tau(v)$. The transgression homomorphism $\mathrm{tr} \colon \mathrm{Hom}_P(T_{\mathbb{Z},n}, A_e) \to H^2(S_n, A_e)$ is defined as

$$\mathrm{tr} \colon f \mapsto f \circ \gamma + B^2(S_n, A_e).$$

b) For $i = 1, 2$ the restriction homomorphism $\mathrm{res} \colon H^i(S_\mathbb{Z}, A_e) \to H^i(T_{\mathbb{Z},n}, A_e)$ is defined as

$$\mathrm{res} \colon \delta + B^i(S_\mathbb{Z}, A_e) \mapsto \delta' + B^i(T_{\mathbb{Z},n}, A_e),$$

where δ' is the restriction of δ to $T_{\mathbb{Z},n}$ and $T_{\mathbb{Z},n} \times T_{\mathbb{Z},n}$, respectively.

c) For $i = 1, 2$ the inflation homomorphism $\mathrm{inf} \colon H^i(S_n, A_e) \to H^i(S_\mathbb{Z}, A_e)$ is defined as

$$\mathrm{inf} \colon \delta + B^i(S_n, A_e) \mapsto \delta' + B^i(S_\mathbb{Z}, A_e),$$

where $\delta' = \delta \circ \pi$ and $\delta' = \delta \circ (\pi, \pi)$, respectively, with $\pi \colon S_\mathbb{Z} \to S_n$ the projection.

These homomorphisms clearly depend on the groups S_n and A_e. However, we omit corresponding labels whenever domain and codomain of these mappings follow from the context.

6.20 Theorem. *The sequence* $0 \to H^1(T_{\mathbb{Z},n}, A_e)^P \xrightarrow{\text{tr}} H^2(S_n, A_e) \xrightarrow{\text{inf}} H^2(S_{\mathbb{Z}}, A_e)$ *is exact.*

Proof. By [66, Theorem 11.9.1], the Hochschild-Serre spectral sequence yields an exact sequence

$$0 \to H^1(S_n, A_e) \xrightarrow{\text{inf}} H^1(S_{\mathbb{Z}}, A_e) \xrightarrow{\text{res}} H^1(T_{\mathbb{Z},n}, A_e)^P \xrightarrow{\text{tr}} H^2(S_n, A_e) \xrightarrow{\text{inf}} H^2(S_{\mathbb{Z}}, A_e).$$

It follows from Lemma 6.18 that inf: $H^1(S_n, A_e) \to H^1(S_{\mathbb{Z}}, A_e)$ is an isomorphism, that is, res = 0, which proves the theorem. □

The following corollary is a consequence of Lemmas 6.18 and 6.17 and Theorem 6.20.

6.21 Corollary. *The order of* $H^2(S_n, A_e)$ *is at most* $|A_e^P|^3 |A_e|^{d/2} = p^{ed/2+3}$.

6.5.3 The second cohomology of S_n

Again, let $n \geq p+1$ and $1 \leq e \leq \mathfrak{e}_n$ in this paragraph. We define certain subgroups of $H^2(S_n, A_e)$ and, as a first step, we determine the group of coboundary tail vectors.

6.22 Lemma. *The tail vector* $x = (x_{i,j})$ *of* S_n *in* A_e *is a coboundary tail vector if and only if* $x_{0,1}, \ldots, x_{0,d-1} \in A_e$ *and* $x_{0,d} = \prod_{k=1}^{d-1} x_{0,k}^{-\beta_k}$ *with* $\beta_k = 1 + \mathfrak{g} + \ldots + \mathfrak{g}^{d-k}$, *and all other tails are trivial; that is,*

$$\mathcal{B}(S_n, A_e) \cong A_e^{d-1}.$$

Proof. Consistency checks as in the proof of Lemma 6.17 show that every coboundary tail vector of S_n in A_e is of the type as in the lemma and, conversely, every list of tails as given in the lemma lies in $\mathcal{B}(S_n, A_e)$. For these consistency checks, note that $\mathfrak{t}_d^{\mathfrak{g}} = \mathfrak{t}_1^{l_{n,1}} \ldots \mathfrak{t}_d^{l_{n,d}}$ in S_n and $a^{l_{n,j}} = a^{-1}$ in A_e for all $a \in A_e$ and $1 \leq j \leq d$, see Remark 6.9 and Lemma 6.1. □

For a P-homomorphism $f: T/T_n \wedge T/T_n \to A_e$ let the 2-cocycle $\Gamma_f: S_n \times S_n \to A_e$ be defined as

$$\Gamma_f: (\mathfrak{g}^i a, \mathfrak{g}^j b) \mapsto f(a^{\mathfrak{g}^j} \wedge b)^{1/2} \quad (a, b \in T/T_n).$$

If $\Gamma_f \in B^2(S_n, A)$ and $a, b \in T/T_n$, then $\Gamma_f(a \wedge b) = \Gamma_f(b \wedge a)$ and, thus, $f(a \wedge b) = 1$. Hence, there is an embedding

$$\text{Hom}_P(T/T_n \wedge T/T_n, A_e) \to H^2(S_n, A_e), \quad f \mapsto \Gamma_f + B^2(S_n, A_e).$$

We now use Lemma 5.8 and consider a P-homomorphism $T \wedge T \to A_e$ as a P-homomorphism $T/T_n \wedge T/T_n \to A_e$, and vice versa.

6.23 Definition. a) If $f: T \wedge T \to A_e$ is a P-homomorphism, then $x_f = x_{\Gamma_f}$ denotes the tail vector of S_n in A_e defined by the 2-cocycle Γ_f, cf. Remark 6.15.

b) The group

$$\text{H}(n, e) = \{x_f \mid f \in \text{Hom}_P(T \wedge T, A_e)\}$$

consists of the hom tail vectors of S_n in A_e.

6.24 Lemma. *The hom tail vector* $x_f = (x_{i,j})$ *satisfies* $x_{i,j} = 1$ *if* $i = 0$ *or* $i = j$, *and* $x_{i,j} = f(\mathfrak{t}_j \wedge \mathfrak{t}_i)$ *if* $1 \leq i < j \leq d$. *The group of hom tail vectors is of the type*

$$\text{H}(n, e) \cong C_p \times A_e^{(p-3)/2}.$$

Proof. This follows from the definition, cf. Remark 6.15 and Corollary 5.6. □

6.5. Cohomology

By Theorem 6.20, the transgression homomorphism

$$\mathrm{tr}\colon \mathrm{Hom}_P(T_{\mathbb{Z},n}, A_e) \to H^2(S_n, A_e), \quad f \mapsto f\circ\gamma + B^2(S_n, A_e),$$

is an embedding where $\gamma \in Z^2(S_n, T_{\mathbb{Z},n})$ is defined by $\gamma(g,h) = \tau(gh)^{-1}\tau(g)\tau(h)$ and $\tau\colon S_n \to S_{\mathbb{Z}}$ is the canonical transversal with $\tau(s) = s$. Recall that $x_{f\circ\gamma}$ is the tail vector defined by the 2-cocycle $f\circ\gamma$, cf. Remark 6.15.

Definition. The group of mainline tail vectors of S_n in A_e is defined as

$$\mathrm{M}(n,e) = \{x_{f\circ\gamma} \mid f \in \mathrm{Hom}_P(T_{\mathbb{Z},n}, A_e)\}.$$

6.25 Lemma. *The mainline tail vector $x_{f\circ\gamma} = (x_{i,j})$ satisfies $x_{i,j} = 1$ if $1 \leq i < j \leq d$ or if $i = 0$ and $j < d$. If $x_{1,1}, \ldots, x_{d,d}$ are trivial, then $x_{0,d}$ is trivial as well. The group of mainline tail vectors is of the type*

$$\mathrm{M}(n,e) \cong A_e.$$

Proof. By Theorem 6.20, the group $\mathrm{M}(n,e)$ is isomorphic to $\mathrm{Hom}_P(T_{\mathbb{Z},n}, A_e) \cong A_e$. Let $x = (x_{i,j})$ be the mainline tail vector defined by $f \in \mathrm{Hom}_P(T_{\mathbb{Z},n}, A_e)$. By Remark 6.15, the tail $x_{i,j}$ is trivial if $1 \leq i < j \leq d$ or if $i = 0$ and $j < d$. If $n = \mathfrak{x}d + \mathfrak{i}$ with integers $\mathfrak{x} \geq 1$ and $1 \leq \mathfrak{i} \leq d$, then

$$x_{i,i} = f(w_i) \quad \text{with} \quad w_i = t_i^{p^{\mathfrak{x}}} w_{n,i}^{-1} \in T_{\mathbb{Z},n} \quad \text{if } 1 \leq i \leq d-\mathfrak{i}+1 \text{ and}$$

$$x_{j,j} = f(w_j) \quad \text{with} \quad w_j = t_j^{p^{\mathfrak{x}+1}} \in T_{\mathbb{Z},n} \quad \text{if } d-\mathfrak{i}+2 \leq j \leq d,$$

where the inverses of $w_{n,1}, \ldots, w_{n,d-\mathfrak{i}+1}$ are computed in $T_{\mathbb{Z}}$. The tail $x_{0,d}$ is

$$x_{0,d} = f(w) \quad \text{with} \quad w = \mathfrak{t}_1^{-l_{n,1}-1} \ldots \mathfrak{t}_d^{-l_{n,d}-1} \in T_{\mathbb{Z},n},$$

and a straightforward, but technical, computation shows that $w = w_1^{-1} \ldots w_d^{-1} u$ for some word u in $\{w_{d-\mathfrak{i}+2}, \ldots, w_d\}$; for a detailed proof we refer to Section A.1.4. This implies that $x_{0,d} = x_{1,1}^{-1} \ldots x_{d,d}^{-1} f(u)$ which proves the lemma. □

Definition. The group of twig tail vectors of S_n in A_e is defined as

$$\mathrm{T}(n,e) = \{(x_{i,j}) \mid x_{0,0}, x_{0,d} \in A_e^P \text{ and } x_{i,j} = 1 \text{ if } (i,j) \notin \{(0,0),(0,d)\}\}.$$

Consistency checks as in the proof of Lemma 6.17 prove the following lemma.

Lemma. *The elements of $\mathrm{T}(n,e)$ are tail vectors of S_n in A_e, and $\mathrm{T}(n,e) \cong C_p^2$.*

We have defined four subgroups of $\mathcal{Z}(S_n, A_e)$ and the following theorem shows that these subgroups suffice to describe $\mathcal{Z}(S_n, A_e)$ completely.

6.26 Theorem. *The group $\mathcal{Z}(S_n, A_e)$ of tail vectors of S_n in A_e can be written as*

$$\mathcal{Z}(S_n, A_e) = \mathcal{B}(S_n, A_e) \oplus \mathrm{M}(n,e) \oplus \mathrm{H}(n,e) \oplus \mathrm{T}(n,e),$$

and so

$$H^2(S_n, A_e) \cong C_p^3 \times A_e^{(p-1)/2}.$$

Proof. By construction, the sum of the right side of the equation is a direct sum and a subgroup of $\mathcal{Z}(S_n, A_e)$ of order $p^{e(3d/2-1)+3}$. By Corollary 6.21 and Lemmas 4.7 and 6.22, the order of $\mathcal{Z}(S_n, A_e)$ is at most $p^{e(3d/2-1)+3}$. □

6.27 Remark. The results of the previous paragraphs show that the exact sequence of Theorem 6.20 splits; that is,

$$H^2(S_n, A_e) \cong H^1(T_{\mathbb{Z},n}, A_e)^P \times H^2(S_{\mathbb{Z}}, A_e),$$

and a splitting $H^2(S_{\mathbb{Z}}, A_e) \to H^2(S_n, A_e)$ in terms of tail vectors is given by $(x_{i,j}) \mapsto (y_{i,j})$ where $y_{i,j} = x_{i,j}$ if $(i,j) \notin \{(1,1), \ldots, (d,d)\}$ and $y_{j,j} = 1$ for all $1 \leq j \leq d$.

6.5.4 Notation

First, we introduce a set of canonical coset representatives of $\mathcal{B}(S_n, A_e)$ in $\mathcal{Z}(S_n, A_e)$.

6.28 Remark. It is well-known that two 2-cocycles which differ by a 2-coboundary define isomorphic extensions, and the same holds for two tail vectors which differ by a coboundary tail vector. Thus, we consider $\mathcal{Z}(S_n, A_e)/\mathcal{B}(S_n, A_e)$ and, from now on, we identify

$$\mathcal{Z}(S_n, A_e)/\mathcal{B}(S_n, A_e) = \mathrm{M}(n, e) \oplus \mathrm{H}(n, e) \oplus \mathrm{T}(n, e),$$

where the direct sum on the right side is a complement to $\mathcal{B}(S_n, A_e)$ in $\mathcal{Z}(S_n, A_e)$, see Theorem 6.26. In particular, we assume that every tail vector $(x_{i,j})$ of S_n in A_e has trivial coboundary tails $x_{0,1}, \ldots, x_{0,d-1}$. Hence, if G is a maximal class group at depth e in the body $\mathcal{T}_n$, then there exists a tail vector $x \in \mathrm{M}(n,e) \oplus \mathrm{H}(n,e) \oplus \mathrm{T}(n,e)$ such that $G \cong E(x)$.

Following the convention made in Remark 6.28, every tail vector x of S_n in A_e can be written uniquely as

$$x = t + h + m$$

where $t \in \mathrm{T}(n,e)$, $h \in \mathrm{H}(n,e)$, and $m \in \mathrm{M}(n,e)$. We call t, h, and m the **twig**, **hom**, and **mainline component of** x, respectively.

The aim of the following chapter is to determine the tail vectors of S_n in A_e which define the groups in the body $\mathcal{T}_n$. For this discussion, a more subtle notation is required. In particular, the following subgroups and subsets of $\mathcal{Z}(S_n, A_e)$ play an important role.

6.29 Definition. a) Recall that the group of hom tail vectors $\mathrm{H}(n,e)$ consists of the tail vectors x_f with $f \in \mathrm{Hom}_P(T \wedge T, A_e)$. We now define

$$\widehat{\mathrm{H}}(n,e) = \{x_f \in \mathrm{H}(n,e) \mid f \text{ surjective}\},$$
$$\mathrm{L}(n,e) = \{x_f \in \mathrm{H}(n,e) \mid f \text{ liftable to a } P\text{-homomorphism } T \wedge T \to T\}, \text{ and}$$
$$\widehat{\mathrm{L}}(n,e) = \widehat{\mathrm{H}}(n,e) \cap \mathrm{L}(n,e).$$

b) Adding the group of twig tail vectors, we define

$$\mathrm{TH}(n,e) = \mathrm{T}(n,e) \oplus \mathrm{H}(n,e) \text{ and}$$
$$\widehat{\mathrm{TH}}(n,e) = \mathrm{T}(n,e) \oplus \widehat{\mathrm{H}}(n,e) \text{ as sets if } e \geq 2, \text{ and, } \widehat{\mathrm{TH}}(n,1) = \mathrm{TH}(n,1) \setminus \{0\}.$$

Using Remark 6.28, these subsets are considered as subsets of $\mathcal{Z}(S_n, A_e)/\mathcal{B}(S_n, A_e)$.

6.5. Cohomology

The following corollary is a consequence of Corollaries 5.6 and 5.7.

6.30 Corollary. *The group* $H(n, e)$ *of hom tail vectors can be written as*

$$H(n, e) = L(n, e) \oplus H_t(n, e)$$

where $L(n, e) \cong A_e^{d/2-1}$ *and* $H_t(n, e) \cong C_p$ *is generated by* $x_{\widehat{f}_2}$, *see Definition 5.5.*

The subscript 't' indicates that the elements of $H_t(n, e)$ have similar properties as the twig tail vectors of S_n in A_e; we comment on these properties the following chapter. For $e \geq 2$, Figure 6.1 depicts the subset and superset relations between the subgroups and subsets defined in this paragraph.

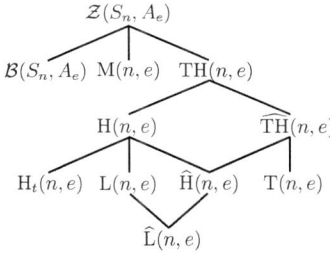

Figure 6.1: Subgroups and subsets of $\mathcal{Z}(S_n, A_e)$.

7 Cohomological description of maximal class groups

The aim of this chapter is to describe the groups in the bodies $\mathcal{T}_{p+1}, \mathcal{T}_{p+2}, \ldots$ of the coclass tree of $\mathcal{G}(p)$ using cohomology. It is proved in Lemma 6.6 that every group at depth e in the body $\mathcal{T}_n$ is an extension of the S-module A_e by the mainline group S_n, and these extensions can be described by elements of the second cohomology group $H^2(S_n, A_e)$. We have seen that
$$H^2(S_n, A_e) \cong \mathcal{Z}(S_n, A_e)/\mathcal{B}(S_n, A_e) = \text{T}(n,e) \oplus \text{H}(n,e) \oplus \text{M}(n,e)$$
for $n \geq p+1$ and $1 \leq e \leq \mathfrak{e}_n$, where the equation on the right side follows from the conventions made in Remark 6.28. The representation of the second cohomology group as a group of tail vectors allows an explicit construction of the extensions defined by their elements: For a tail vector x of S_n in A_e the corresponding extension $E(x)$ is the group defined by the presentation $\mathcal{E}(x)$, see Definition 6.14. The definition of this presentation depends on the standard p.c.p.s of the groups S_n and A_e, see Section 6.3, and the elements of $E(x)$ are normalized words in $\mathcal{S} \cup A_e$ where $\mathcal{S} = \{\mathfrak{g}, \mathfrak{t}_1, \ldots, \mathfrak{t}_d\}$.

In this chapter, we characterize the tail vectors of S_n in A_e which define the groups in the bodies of the coclass tree. More precisely, we describe a set of tail vectors of S_n in A_e which define, up to isomorphism, all the groups at depth e in the body $\mathcal{T}_n$. Thus, as a first step, we characterize the tail vectors defining the groups in question. The second step is then to discuss the isomorphism problem, that is, the construction of the groups up to isomorphism.

Unless otherwise noted, let $n \geq p+1$ and $1 \leq e \leq \mathfrak{e}_n$ throughout this chapter.

7.1 Tail vectors defining maximal class groups

Let x be a tail vector of S_n in A_e. Recall that x can be written uniquely as $x = t + h + m$ where t, h, and m are the twig, hom, and mainline component of x, respectively. If $u, v \in E(x)$ with $u = v$ in $E(x)$, then we also say that $E(x)$ **satisfies** $u = v$.

We start with two preliminary lemmas.

7.1 Lemma. *Let x be a tail vector of S_n in A_e with twig component $t = (t_{i,j})$. Then $E(x)$ satisfies*
$$\mathfrak{g}^p = t_{0,0} \quad \text{and} \quad (\mathfrak{g}\mathfrak{t}_1)^p = t_{0,0} t_{0,d}.$$

Proof. Let $m = (m_{i,j})$ be the mainline component of x. Recall that $\mathfrak{t}_d^{\mathfrak{g}} = \mathfrak{t}_1^{l_{n,1}} \ldots \mathfrak{t}_d^{l_{n,d}}$ in S_n and $l_{n,1}, \ldots, l_{n,d}$ are all congruent to -1 modulo the exponent of A_e, see Remark 6.9 and Lemma 6.1. This shows that $E(x)$ satisfies
$$\mathfrak{g}^p = t_{0,0} \quad \text{and} \quad (\mathfrak{g}\mathfrak{t}_1)^p = \mathfrak{t}_1^{l_{n,1}+1} \ldots \mathfrak{t}_d^{l_{n,d}+1} m_{0,d} t_{0,0} t_{0,d}.$$

Hence, the lemma is proved if $(\mathfrak{g}\mathfrak{t}_1)^p = 1$ in $E(m)$. This follows from Lemma 6.25 and Remark 6.28 and a straightforward, but technical, computation. For a detailed proof we refer to Section A.1.4. □

If $\pi\colon A_e \to A_i$ with $i \leq e$ is the projection, then we apply π to a tail vector of S_n in A_e by applying it to the corresponding tails of the vector. Recall that a maximal class group is capable if it has an immediate descendant in the coclass graph $\mathcal{G}(p)$.

7.2 Lemma. *Let $x \in \mathcal{Z}(S_n, A_e)$ such that $E(x)$ has maximal class.*
a) *If $i \leq e$ and $\pi\colon A_e \to A_i$ is the projection, then $E(\pi(x)) \cong E(x)/\gamma_{n+i}(E(x))$.*
b) *If $E(x)$ is capable, then $\mathfrak{g}^p = (\mathfrak{g}\mathfrak{t}_1)^p = 1$ in $E(x)$.*

Proof. a) By construction, $\pi(x)$ is the tail vector defined by the 2-cocycle $\pi \circ \gamma_x$, see Definition 4.8, and $E(x) \cong E(\gamma_x)$ and $E(\pi(x)) \cong E(\pi \circ \gamma_x)$. Clearly, the projection $E(\gamma_x) \to E(\pi \circ \gamma_x)$ has kernel $\gamma_{n+i}(E(\gamma_x))$.
b) Let G be an immediate descendant of $E(x)$ and let $g, t_1 \in G$ be preimages of $\mathfrak{g}, \mathfrak{t}_1 \in E(x)$ under the projection $G \to E(x)$. It follows from Lemma 3.3 that g^p and $(gt_1)^p$ lie in the kernel of this projection, that is, $\mathfrak{g}^p = (\mathfrak{g}\mathfrak{t}_1)^p = 1$ in $E(x)$. $\square$

7.1.1 Mainline tail vectors

In this paragraph, we consider the tail vectors of S_n in A_e which define extensions isomorphic to the mainline group S_{n+e}. The following lemma motivates the notation of the group of mainline tail vectors.

7.3 Lemma. *Let x be a tail vector of S_n in A_e such that $E(x)$ has maximal class. Then $E(x) \cong S_{n+e}$ if and only if $x \in \mathrm{M}(n, e)$.*

Proof. If $E(x) \cong S_{n+e}$, then the 2-step centralizer $P_1(E(x))$ is abelian and $\mathfrak{g}^p = (\mathfrak{g}\mathfrak{t}_1)^p = 1$ in $E(x)$ by Lemma 7.2; that is, x has trivial tails $x_{i,j}$ for $1 \leq i < j \leq d$, and $x \in \mathrm{M}(n, e)$ by Lemma 7.1. Conversely, if $x \in \mathrm{M}(n, e)$ and $E(x)$ has maximal class, then $P_1(E(x))$ is abelian and $E(x)$ satisfies $\mathfrak{g}^p = (\mathfrak{g}\mathfrak{t}_1)^p = 1$ by Lemma 7.1; that is, $E(x) \cong S_{n+e}$ by Corollary 6.5. $\square$

We now construct an explicit mainline tail vector of S_n in A_e which defines S_{n+e}. For this purpose, let $\tau\colon S_n \to S_{n+e}$ be the canonical transversal with $\tau(s) = s$ and define

$$\gamma_{n,e}\colon S_n \times S_n \to T_n/T_{n+e}, \quad (u,v) \mapsto \tau(uv)^{-1}\tau(u)\tau(v),$$

and

$$\gamma'_{n,e} = \iota_{n,e} \circ \gamma_{n,e} \in Z^2(S_n, A_e),$$

where $\iota_{n,e}\colon T_n/T_{n+e} \to A_e$ is the P-isomorphism of Definition 6.2. Recall that $x_{\gamma'_{n,e}}$ is the tail vector of S_n in A_e defined by the 2-cocycle $\gamma'_{n,e}$, cf. Remark 6.15.

Definition. The mainline tail vector $\mathfrak{m}_{n,e}$ of S_n in A_e is defined as $\mathfrak{m}_{n,e} = x_{\gamma'_{n,e}}$.

7.4 Lemma. a) *The groups S_{n+e}, $E(\gamma_{n,e})$, and $E(\mathfrak{m}_{n,e})$ are isomorphic.*
b) *The extensions $E(\gamma_{n,e})$ and S_{n+e} are equivalent.*
c) *As tail vectors, $\mathfrak{m}_{n,e} = \mathfrak{m}_{n+d,e}$, and $\mathfrak{m}_{n,e} \in \mathrm{M}(n, e)$.*
d) *If $e \geq 1$ and $\pi\colon A_e \to A_{e-1}$ is the projection, then $\pi(\mathfrak{m}_{n,e}) = \mathfrak{m}_{n,e-1}$.*

Proof. a) This follows from the construction and Remark 6.16.
b) The isomorphism $E(\gamma_{n,e}) \to S_{n+e}$, $(u, a) \mapsto \tau(u)a$, proves the equivalence.

7.1. Tail vectors defining maximal class groups

c) Lemma 7.3 and part a) show that $\mathfrak{m}_{n,e} \in M(n,e)$. It follows from Remarks 6.9 and 6.15 and the construction of the 2-cocycles $\gamma'_{n,e}$ and $\gamma'_{n+d,e}$ that $\mathfrak{m}_{n,e} = \mathfrak{m}_{n+d,e}$.

d) The tail vector $\pi(\mathfrak{m}_{n,e})$ is defined by $\pi \circ \gamma'_{n,e}$. If $\pi' \colon S_{n+e} \to S_{n+e-1}$ is the projection, then $\pi \circ \gamma'_{n,e} = \iota_{n,e-1} \circ \pi' \circ \gamma_{n,e}$. For $m \geq 0$ let the canonical transversal $S_n \to S_{n+m}$ be denoted by τ_m. Then $\pi' \circ \tau_e = \tau_{e-1}$ implies that $\pi' \circ \gamma_{n,e} = \gamma_{n,e-1}$. □

7.1.2 The characterization of tail vectors

Again, we write a tail vector x of S_n in A_e as $x = t + h + m$ where $t \in T(n,e)$ is a twig tail vector, $h \in H(n,e)$ is a hom tail vector, and $m \in M(n,e)$ is a mainline tail vector. We use this notation in this paragraph.

First, we show that the mainline component of a tail vector decides whether the corresponding extension has maximal class. Part (2) of the proof of the following theorem is motivated by [29, Lemma 8.1.3].

7.5 Theorem. *If $x = t + h + m$ is a tail vector of S_n in A_e, then $E(x)$ has maximal class if and only if $E(m)$ has maximal class.*

Proof. Let y be a tail vector of S_n in A_e. It is sufficient to prove the following.

(1) If $E(y)$ has maximal class, then $E(y+t)$ has maximal class.

(2) If $E(y)$ has maximal class, then $E(y+h)$ has maximal class

Ad (1): Let $s = \mathfrak{g}$ and $s_1 = \mathfrak{t}_1$ in $E(y)$, and iteratively define $s_{j+1} = [s_j, s]$. Since $E(y)$ has maximal class, it follows from Lemma 3.3 that $s_{n+e-1} \neq 1$ in $E(y)$. Note that $E(y) = E(y+t)$ as sets, and the same construction with $s' = \mathfrak{g}$ and $s'_1 = \mathfrak{t}_1$ in $E(y+t)$ yields elements $s', s'_1, \ldots$ in $E(y+t)$. By definition, the tails of t are central and have order dividing p, and, as $n \geq p+1$, the exponents of $\mathfrak{t}_d$ in $s_1, \ldots, s_d$ are all divisible by p, cf. Definition 6.14. Thus, by construction, $s'_j = s_j$ as words for $1 \leq j \leq d$ and there is a central $a \in E(y)$ such that the collection of $s_p a$ in $E(y)$ yields a word equal to s'_p. The exponents of $\mathfrak{t}_1, \ldots, \mathfrak{t}_d$ in s_p are divisible by p and, hence, $s'_j = s_j$ as words for all $j > p$. This shows that $s'_{n+e-1} \neq 1$ in $E(y+t)$, and $E(y+t)$ has maximal class.

Ad (2): Recall that h is the tail vector defined by a 2-cocycle Γ_f for some P-homomorphism $f \colon T \wedge T \to A_e$, see Definition 6.23, and let $\gamma_y \colon S_n \times S_n \to A_e$ be the canonical 2-cocycle defined by y, see Definition 4.8. We use restrictions of these cocycles to define the extensions

$$\widetilde{M} < M < E(\gamma_y) \quad \text{and} \quad \widetilde{N} < N < E(\gamma_y + \Gamma_f)$$

of A_e by T_{n-1}/T_n and T/T_n, respectively. By definition, $E(\gamma_y) = E(\gamma_y + \Gamma_f)$ as sets. By part (1), we can assume that y has a trivial twig component and, thus, M and N both have a complement $P = C_p(\mathfrak{g})$. The restriction of Γ_f to $T_{n-1}/T_n \times T_{n-1}/T_n$ is trivial and, therefore, $\widetilde{M}$ and $\widetilde{N}$ are isomorphic as P-modules. It follows from Lemma 3.5b) that P acts uniserially on $\widetilde{M}$ and, hence, on $\widetilde{N}$. Lemma 3.5a) shows that $E(\gamma_y + \Gamma_f)$ acts uniserially on $\widetilde{N}$ and, by construction, on N/A_e. Now Lemma 3.6 proves that $E(\gamma_y + \Gamma_f)$ acts uniserially on N and, by Lemma 3.4, the group $E(\gamma_y + \Gamma_f)$ has maximal class. □

We now provide a necessary and sufficient condition for a tail vector of S_n in A_e to define an extension in the body $\mathcal{T}_n$. Recall the definition of the set $\widehat{TH}(n,e)$, see Definition 6.29.

7.6 Corollary. *Let $x = t + h + m$ be a tail vector of S_n in A_e. Then $E(x)$ lies in the body T_n if and only if the tail vector m defines a group of maximal class and $t + h \in \widehat{\mathrm{TH}}(n, e)$.*

Proof. By Theorem 7.5, the tail vector x defines a group of maximal class if and only if m does. Let $E(x)$ be of maximal class. Clearly, the group $E(x)$ lies in T_n if and only if S_{n+1} is not a quotient of $E(x)$. Hence, Lemma 7.3 proves the case $e = 1$. We now consider $e > 1$ and the projection $\pi \colon A_e \to A_1$. Then S_{n+1} is not a quotient of $E(x)$ if and only if $E(\pi(x))$ is not isomorphic to S_{n+1}, if and only if $\pi(x) \notin \mathrm{M}(n, 1)$, if and only if $\pi(h) \neq 0$, if and only if the homomorphism $T \wedge T \to A_e$ defining h is surjective, see Lemma 5.9. $\square$

7.2 Isomorphism problem

Every group at depth e in the body T_n is isomorphic to a group defined by a tail vector $x = t + h + m$ of S_n in A_e where m defines S_{n+e} and $t + h$ lies in $\widehat{\mathrm{TH}}(n, e)$. In this section, we attack the isomorphism problem and develop a criterion when two maximal class extension defined by tail vectors are isomorphic.

7.2.1 Compatible pairs

Our main tool for solving the isomorphism problem is the group of compatible pairs, and, as a first step, we recall some definitions made in Section 4.2.1.

The group of compatible pairs of S_n and A_e is defined as

$$\mathrm{Comp}(n, e) = \{(\alpha, \beta) \in \mathrm{Aut}(S_n) \times \mathrm{Aut}(A_e) \mid \forall a \in A_e : \beta(a)^{\alpha(\mathfrak{g})} = \beta(a^{\mathfrak{g}})\}$$

and acts on the group of 2-cocycles $Z^2(S_n, A_e)$ via

$$\gamma^{(\alpha, \beta)} = \beta^{-1} \circ \gamma \circ (\alpha, \alpha).$$

The group $B^2(G, N)$ of 2-coboundaries is invariant under this action and, hence, the group of compatible pairs acts on $H^2(G, N)$. If $\gamma, \delta \in H^2(S_n, A_e)$ are cohomology classes, then the extensions $E(\delta)$ and $E(\gamma)$ are strongly isomorphic if and only if $\gamma^c = \delta$ for some compatible pair $c \in \mathrm{Comp}(n, e)$, see Theorem 4.3. In general, there are extensions which are isomorphic but not strongly isomorphic. However, in the case of maximal class extensions, the action of the compatible pairs can be used to obtain a reduction up to isomorphism, see [14, Theorem 20].

7.7 Lemma. *If γ and δ are cohomology classes in $H^2(S_n, A_e)$ defining groups of maximal class, then $E(\gamma) \cong E(\delta)$ if and only if $\gamma^c = \delta$ for some compatible pair $c \in \mathrm{Comp}(n, e)$.*

Proof. If γ and δ define groups of maximal class, then A_e is the n-th term of the lower central series of the corresponding group extensions. Thus, every isomorphism from $E(\delta)$ to $E(\gamma)$ is a strong isomorphism and Theorem 4.3 proves the lemma. $\square$

Recall that $H^2(S_n, A_e) \cong \mathcal{Z}(S_n, A_e)/\mathcal{B}(S_n, A_e)$ is identified with

$$\mathrm{T}(n, e) \oplus \mathrm{H}(n, e) \oplus \mathrm{M}(n, e),$$

and we assume that every tail vector of S_n in A_e has trivial coboundary tails. We now show how a compatible pair acts on this group.

7.2. Isomorphism problem

Definition. If $c = (\alpha, \beta)$ is a compatible pair of S_n and A_e and if x is a tail vector of S_n in A_e, then x^c is the tail vector defined by the 2-cocycle

$$\gamma^c = \beta^{-1} \circ \gamma \circ (\alpha, \alpha),$$

where $\gamma \colon S_n \times S_n \to A_e$ is any 2-cocycle defining x, for example $\gamma = \gamma_x$ as in Definition 4.8. Recall that x^c is modified by a suitable coboundary tail vector so that x^c has trivial coboundary tails.

The group of compatible pairs of S and A_e is denoted by

$$\mathrm{Comp}(e) = \{(\alpha, \beta) \in \mathrm{Aut}(S) \times \mathrm{Aut}(A_e) \mid \forall a \in A_e : \beta(a)^{\alpha(\mathfrak{g})} = \beta(a^{\mathfrak{g}})\}$$

and, by Theorem 6.13, the mapping

$$\mathrm{Comp}(e) \to \mathrm{Comp}(n, e), \quad (\alpha, \beta) \mapsto (\alpha|_{S_n}, \beta),$$

is surjective. This furnishes $\mathcal{Z}(S_n, A_e)/\mathcal{B}(S_n, A_e)$ with the structure of a $\mathrm{Comp}(e)$-module, and the following corollary transfers Lemma 7.7 to extensions defined by tail vectors.

7.8 Corollary. *If x and y are tail vectors of S_n in A_e defining extensions of maximal class, then $E(x) \cong E(y)$ if and only if $x^c = y$ for some compatible pair $c \in \mathrm{Comp}(e)$.*

Proof. The groups $E(x)$ and $E(y)$ are isomorphic if and only if $E(\gamma_x) \cong E(\gamma_y)$, if and only if there is $c \in \mathrm{Comp}(n, e)$ with $\gamma_x^c \equiv \gamma_y \mod \mathcal{B}^2(S_n, A_e)$, if and only if $x^c = y$. □

Lemma 7.3 and Corollary 7.8 show that the tail vectors of S_n in A_e which define the mainline group S_{n+e} all lie in the same $\mathrm{Comp}(e)$-orbit as the mainline tail vector $\mathfrak{m}_{n,e}$.

7.9 Corollary. *If $x \in \mathrm{M}(n, e)$ defines a group of maximal class, then $x^c = \mathfrak{m}_{n,e}$ for some compatible pair $c \in \mathrm{Comp}(e)$.*

We now reconsider some subsets of $\mathcal{Z}(S_n, A_e)/\mathcal{B}(S_n, A_e)$ defined in Definition 6.29 and show that they are invariant under the action of $\mathrm{Comp}(e)$. This is the basis for the proof of the subsequent theorem.

7.10 Lemma. *The sets $\mathrm{TH}(n, e)$, $\mathrm{T}(n, e)$, and $\widehat{\mathrm{TH}}(n, e)$ are $\mathrm{Comp}(e)$-invariant.*

Proof. Let $c = (\alpha, \beta)$ be an element of $\mathrm{Comp}(e)$. First, we consider a hom tail vector $x_f \in \mathrm{H}(n, e)$ of S_n in A_e defined by a P-homomorphism $f \colon T \wedge T \to A_e$. We define

$$f^c = \beta^{-1} \circ f \circ (\alpha \wedge \alpha)|_{T \wedge T}$$

and note that f^c is a P-homomorphism as well. By construction, x_f^c and x_{f^c} are the tail vectors defined by the 2-cocycles $(\Gamma_f)^c$ and Γ_{f^c}, respectively, and

$$(\Gamma_f)^c|_{T/T_n \times T/T_n} = \Gamma_{f^c}|_{T/T_n \times T/T_n}.$$

Hence, it follows from Remark 6.15 that $t = (x_f)^c - x_{f^c}$ has trivial tails $t_{1,1}, \ldots, t_{d,d}$ and $t_{i,j}$ for $1 \leq i < j \leq d$; that is, t is a twig tail vector and $x_f^c \in \mathrm{TH}(n, e)$.

Analogously, it is straightforward to prove that $x^c \in \mathrm{T}(n, e)$ for all twig tail vectors $x \in \mathrm{T}(n, e)$ of S_n in A_e. If $f \colon T \wedge T \to A_e$ is a surjective P-homomorphism, then f^c is surjective as well. Again, $(x_f)^c - x_{f^c} \in \mathrm{T}(n, e)$, which proves the lemma. □

The main result of this paragraph is given in the following theorem.

7.11 Theorem. *A complete set of isomorphism types of groups at depth e in the body $\mathcal{T}_n$ is given by*
$$\{E(\mathfrak{m}_{n,e} + y) \mid y \in \widehat{\mathrm{TH}}(n,e)\}.$$
If $x, y \in \widehat{\mathrm{TH}}(n,e)$, then
$$E(\mathfrak{m}_{n,e} + x) \cong E(\mathfrak{m}_{n,e} + y)$$
if and only if $x^c = y$ for some $c \in \mathrm{Stab}_{\mathrm{Comp}(e)}(\mathfrak{m}_{n,e})$.

Proof. Let G be a group at depth e in the body $\mathcal{T}_n$. By Remark 6.28, there is a tail vector $x = m + y$ with $m \in \mathrm{M}(m,e)$ and $y \in \mathrm{TH}(n,e)$ such that $E(x) \cong G$. Theorem 7.5 and Corollary 7.6 show that m defines the mainline group S_{n+e} and y lies in $\widehat{\mathrm{TH}}(n,e)$. By Corollary 7.9, there is a compatible pair c with $m^c = \mathfrak{m}_{n,e}$; that is, $E(x) \cong E(\mathfrak{m}_{n,e} + y^c)$ with $y^c \in \widehat{\mathrm{TH}}(n,e)$ by Lemma 7.10. By Theorem 7.5 and Corollary 7.6, all groups in the set of the theorem have depth e in the body $\mathcal{T}_n$. Now the assertion follows from Corollary 7.8 and Lemma 7.10. □

7.2.2 Aut(S)-module action

Every group at depth e in the body $\mathcal{T}_n$ is isomorphic to a group $E(\mathfrak{m}_{n,e} + y)$ for some $y \in \widehat{\mathrm{TH}}(n,e)$, and two groups $E(\mathfrak{m}_{n,e}+y)$ and $E(\mathfrak{m}_{n,e}+x)$ with $x, y \in \widehat{\mathrm{TH}}(n,e)$ are isomorphic if and only if $x^c = y$ for some $c \in \mathrm{Stab}_{\mathrm{Comp}(e)}(\mathfrak{m}_{n,e})$. We now consider this stabilizer in more detail and write
$$\Sigma_{n,e} = \mathrm{Stab}_{\mathrm{Comp}(e)}(\mathfrak{m}_{n,e}).$$

As a first step, we use the P-isomorphism $\iota_{n,e} \colon T_n/T_{n+e} \to A_e$ of Definition 6.2 to define a homomorphism from $\mathrm{Aut}(S)$ to $\mathrm{Aut}(A_e)$.

7.12 Definition. If $\alpha \in \mathrm{Aut}(S)$, then $\xi_{n,e}(\alpha) \in \mathrm{Aut}(A_e)$ is defined as
$$\xi_{n,e}(\alpha) = \iota_{n,e} \circ \alpha|_{T_n/T_{n+e}} \circ \iota_{n,e}^{-1}.$$

By definition, if $\alpha \in \mathrm{Aut}(S)$ acts trivially on $P = S/T$, then $\xi_{n,e}(\alpha) = \alpha|_{A_e}$.

7.13 Lemma. *The group $\Sigma_{n,e}$ can be described by*
$$\Sigma_{n,e} = \{(\alpha, \xi_{n,e}(\alpha)) \mid \alpha \in \mathrm{Aut}(S)\}.$$

Proof. We consider the 2-cocycle $\gamma_{n,e} \colon S_n \times S_n \to T_n/T_{n+e}$ of Section 7.1.1 and write $\gamma = \gamma_{n,e} + B^2(S_n, T_n/T_{n+e})$ for the corresponding cohomology class. Recall that $\gamma'_{n,e}$ is defined as $\gamma'_{n,e} = \iota_{n,e} \circ \gamma_{n,e}$. It follows from Lemma 4.4 that
$$\mathrm{Stab}_{\mathrm{Comp}(S_n, T_n/T_{n+e})}(\gamma) = \{(\alpha|_{S_n}, \alpha|_{T_n/T_{n+e}}) \mid \alpha \in \mathrm{Aut}(E(\gamma_{n,e}))\}.$$

By Lemma 7.4, the extensions S_{n+e} and $E(\gamma_{n,e})$ are equivalent which implies that
$$\mathrm{Stab}_{\mathrm{Comp}(S_n, T_n/T_{n+e})}(\gamma) = \{(\alpha|_{S_n}, \alpha|_{T_n/T_{n+e}}) \mid \alpha \in \mathrm{Aut}(S_{n+e})\},$$
and, by Theorem 6.13,
$$\mathrm{Stab}_{\mathrm{Comp}(S_n, T_n/T_{n+e})}(\gamma) = \{(\alpha|_{S_n}, \alpha|_{T_n/T_{n+e}}) \mid \alpha \in \mathrm{Aut}(S)\}.$$

7.3. Twigs, skeleton, and capable groups

The P-isomorphism $\iota_{n,e}$ defines an isomorphism

$$\mathrm{Comp}(S_n, T_n/T_{n+e}) \to \mathrm{Comp}(S_n, A_e), \quad (\alpha, \beta) \to (\alpha, \iota_{n,e} \circ \beta \circ \iota_{n,e}^{-1}),$$

and

$$\beta^{-1} \circ \gamma_{n,e} \circ (\alpha, \alpha) \equiv \gamma_{n,e} \bmod B^2(S_n, T_n/T_{n+e})$$

with $\alpha \in \mathrm{Aut}(S)$ and $\beta \in \mathrm{Aut}(T_n/T_{n+e})$ if and only if

$$\iota_{n,e} \circ \beta^{-1} \circ \iota_{n,e}^{-1} \circ \gamma'_{n,e} \circ (\alpha, \alpha) \equiv \gamma'_{n,e} \bmod B^2(S_n, A_e).$$

This implies that

$$\mathrm{Stab}_{\mathrm{Comp}(e)}(\gamma'_{n,e} + B^2(S_n, A_e)) = \{(\alpha, \xi_{n,e}(\alpha)) \mid \alpha \in \mathrm{Aut}(S)\},$$

which proves the lemma. □

Lemma 7.13 shows that $\mathrm{TH}(n,e)$ can be considered as an $\mathrm{Aut}(S)$-module. We conclude this section with a remark on this action, and we summarize the main results of this section in a corollary.

7.14 Remark. a) The group $\mathrm{Aut}(S)$ acts on $\mathrm{TH}(n,e)$ via

$$\mathrm{Aut}(S) \to \Sigma_{n,e}, \quad \alpha \mapsto (\alpha, \xi_{n,e}(\alpha)).$$

b) The automorphism $\alpha \in \mathrm{Aut}(S)$ acts on $f \in \mathrm{Hom}_P(T \wedge T, A_e)$ via

$$f^\alpha = \xi_{n,e}(\alpha)^{-1} \circ f \circ (\alpha \wedge \alpha)|_{T \wedge T} : T \wedge T \to A_e,$$

and the proof of Lemma 7.10 shows that $(x_f)^\alpha - x_{f^\alpha} \in \mathrm{T}(n,e)$.

7.15 Corollary. *If $\mathcal{M}$ is a set of $\mathrm{Aut}(S)$-orbit representatives in $\widehat{\mathrm{TH}}(n,e)$, then, up to isomorphism, a complete list of groups at depth e in the body $\mathcal{T}_n$ is given by*

$$(E(\mathfrak{m}_{n,e} + y) \mid y \in \mathcal{M}).$$

7.3 Twigs, skeleton, and capable groups

Using the results of the previous sections, we investigate the skeleton and twigs of a body, see Definition 2.3. In particular, we determine the depths of a body and its skeleton, and we show that the twigs are subtrees of depth 1.

7.3.1 Capability

We start with a preliminary lemma. Recall that $\mathrm{L}(n,e) \subseteq \mathrm{H}(n,e)$ is the subset of hom tail vectors which are induced by liftable P-homomorphisms $T \wedge T \to A_e$, see Definition 6.29.

7.16 Lemma. *If $\alpha \in \mathrm{Aut}(S)$ and $x_f \in \mathrm{L}(n,e)$, then $(x_f)^\alpha = x_{f^\alpha}$ lies in $\mathrm{L}(n,e)$.*

Proof. Let $h\colon T \wedge T \to A_{e+1}$ be a lifting of f and let $\pi\colon A_{e+1} \to A_e$ be the projection. By definition, $f^\alpha = \pi \circ h^\alpha$ and f^α is liftable, that is, $x_{f^\alpha} \in \mathrm{L}(n,e)$. It follows from Remark 7.14 that $x_f^\alpha - x_{f^\alpha}$ is a twig tail vector and it remains to show that x_f^α has a trivial twig component. If $e < \mathfrak{e}_n$, then a 2-cocycle inducing x_f is $\Gamma_f = \pi \circ \Gamma_h$, and it follows from $\Gamma_f^\alpha = \pi \circ \Gamma_h^\alpha$ that Γ_f^α induces a tail vector with trivial twig component.

The construction of the 2-cocycle $\Gamma_h : S_n \times S_n \to A_{e+1}$ requires $\mathfrak{e}_n < \lfloor \frac{n-1}{d} \rfloor d$, see Lemma 5.8, and we can use the same argument as above if $e = \mathfrak{e}_n < \lfloor \frac{n-1}{d} \rfloor d$. However, if $e = \mathfrak{e}_n = \lfloor \frac{n-1}{d} \rfloor d$, which is possible only for $p \in \{5,7\}$, then we can define Γ_h in a similar way as a 2-cocycle $S_\mathbb{Z} \times S_\mathbb{Z} \to A_{e+1}$. In this case, we regard Γ_f and Γ_h as 2-cocycles from $S_\mathbb{Z} \times S_\mathbb{Z}$ to A_e and A_{e+1}, respectively. The tail vector x_f^α has a trivial mainline component and, using Remark 6.27, it is induced by $\Gamma_f^\alpha = \pi \circ \Gamma_h^\alpha$. Now the results of Section 6.5.1 imply that x_f^α has a trivial twig component. $\square$

Recall that $\widehat{\mathrm{L}}(n,e) \subseteq \mathrm{L}(n,e)$ is the subset consisting of the hom tail vectors which are induced by surjective P-homomorphisms. The following lemma shows that the capable groups in the body $\mathcal{T}_n$ are basically defined by the elements of $\widehat{\mathrm{L}}(n,e)$.

7.17 Lemma. *If $e < \mathfrak{e}_n$ and $y \in \widehat{\mathrm{TH}}(n,e)$, then $E(\mathfrak{m}_{n,e} + y)$ is capable if and only if y lies in $\widehat{\mathrm{L}}(n,e)$.*

Proof. Let $\pi\colon A_{e+1} \to A_e$ be the projection. First, we assume that $E(x)$ with $x = \mathfrak{m}_{n,e} + y$ is capable, that is, $y = x_f$ for some P-homomorphism $f\colon T \wedge T \to A_e$, see Lemmas 7.2 and 7.1. We can assume that an immediate descendant of $E(x)$ is defined by $x' = \mathfrak{m}_{n,e+1} + x_k + t$ with $t \in \mathrm{T}(n,e+1)$ and $k\colon T \wedge T \to A_{e+1}$. The groups $E(\pi(x'))$ and $E(x)$ are isomorphic and it follows from Theorem 7.11 and Lemmas 7.13, 7.4, and 7.16 that $x_f = \pi(x_k)^\alpha = (x_{\pi \circ k})^\alpha = x_{\pi \circ k^\alpha}$ for some $\alpha \in \mathrm{Aut}(S)$. Thus, f can be lifted to $k^\alpha\colon T \wedge T \to A_{e+1}$ and, therefore, $y \in \widehat{\mathrm{L}}(n,e)$. Conversely, let $x = \mathfrak{m}_{n,e} + x_f$ with $x_f \in \widehat{\mathrm{L}}(n,e)$. If $k\colon T \wedge T \to A_{e+1}$ is a lifting of f, then $E(\mathfrak{m}_{n,e+1} + x_k)$ is an immediate descendant of $E(x)$. $\square$

Lemma 7.17 also motivates the notation of the group $\mathrm{T}(n,e)$. If a tail vector has a nontrivial twig component, then the corresponding group extension is terminal; that is, it is a *twig* in the branch of the coclass tree. The same happens if a tail vector involves a non-trivial element of $\mathrm{H}_t(n,e)$, see Corollary 6.30.

7.3.2 Depths of skeleton, body, and branch

Let $n \geq p+1$ and $e \leq \lfloor \frac{n-1}{d} \rfloor d$ in this paragraph. We start with two preparing lemmas.

7.18 Lemma. *If $f\colon T \wedge T \to T_n$ is a P-homomorphism and $a \in T_n$ and $b \in T_{n-1}$, then $f(a \wedge c) \in T_{n+e}$ and $f(b \wedge c) \in T_{n+d}$ for all $c \in T$.*

Proof. We write $n = \mathfrak{r}d + \mathfrak{i}$ with integers $\mathfrak{r} \geq 1$ and $1 \leq \mathfrak{i} \leq d$. Then $a = t^{p^{\mathfrak{r}}}$ for some $t \in T$ and $f(a \wedge b) = f(t \wedge b)^{p^{\mathfrak{r}}} \in T_{n+\mathfrak{r}d}$. It follows from the choice of e that $T_{n+\mathfrak{r}d} \leq T_{n+e}$. Analogously, $f(b \wedge c) \in T_{n+d}$ follows from $b \in T_p$. $\square$

7.3. Twigs, skeleton, and capable groups

Let $f\colon T\wedge T\to T_n$ be a surjective P-homomorphism and denote by $f'\colon T\wedge T\to T_n/T_{n+e}$ the composition of the projection $T_n\to T_n/T_{n+e}$ and f. It follows from Lemma 7.18 that f' can be applied to elements of $T/T_{n+e}\wedge T/T_{n+e}$. Thus, by Definition 2.2, the skeleton group $C_{f,e}$ defined by f and e is isomorphic to

$$C_{f',e} = P \ltimes (T/T_{n+e}, \odot) \quad \text{with} \quad a\odot b = ab f'(a\wedge b)^{1/2} \quad (a,b\in T/T_{n+e}),$$

and, from now on, we identify $C_{f,e}$ with $C_{f',e}$.

7.19 Lemma. *Let $f\colon T\wedge T\to T_n$ be a surjective P-homomorphism.*

a) *The group $C_{f,e}$ has depth e in the branch $\mathcal{B}_n$.*

b) *If $e < \lfloor \frac{n-1}{d}\rfloor d$, then $C_{f,e+1}$ is a descendant of $C_{f,e}$.*

Proof. a) Let $G = C_{f,e}$ and identify $C_{f,e} = C_{f',e}$ with $f'\colon T/T_n \wedge T/T_n \to T_n/T_{n+e}$ defined as above. By Lemma 7.18, the subgroup $N = T_n/T_{n+e}$ of G is normal and abelian and G acts uniserially on N. Thus, G has maximal class if $\gamma_n(G) = N$. Let s_1 be an element of $(T/T_{n+e}, \odot)$ not lying in T_2/T_{n+e} and, iteratively, define

$$s_{i+1} = [s_i, \mathfrak{g}] = s_i^{\mathfrak{g}-1} f'(s_i \wedge s_i^{\mathfrak{g}})^{-1/2}.$$

It follows from Lemma 7.18 and induction that $s_n \notin T_{n+1}/T_{n+e}$. By construction, s_n lies in $\gamma_n(G)$ and $\gamma_n(G) \leq N$. This implies that $\gamma_n(G) = N$, and G lies in $\mathcal{B}_n$ as f is surjective.

b) This follows from the construction. $\square$

An immediate consequence is the following corollary.

7.20 Corollary. *The skeleton $\mathcal{S}_n$ and body $\mathcal{T}_n$ both have depth $\mathfrak{e}_n$.*

A bound for the depth of a branch is given in the next theorem.

7.21 Theorem. *The depth of $\mathcal{B}_n$ is at most $n + 2p - 11$ if $p \geq 7$ and n if $p = 5$.*

Proof. Let $\mathcal{S}_n \to H_1 \to \ldots \to H_m$ be a path of maximal length in $\mathcal{B}_n$; that is, H_j has order p^{n+j} and positive degree of commutativity l_j for $1 \leq j \leq m$. Assuming that $m > 1$, it follows from Corollary 6.5 that the 2-step centralizer of H_1 is non-abelian. Thus, if H_1 has refined central series $H_1 > P_1 > \ldots > P_{n+1} = \{1\}$, then $[P_1, P_1] = [P_1, P_2] \neq \{1\}$, which implies that $l_1 \leq n-3$ and $l_2, \ldots, l_m \leq l_1$. If $p \geq 7$, then Theorem 3.2 shows that $2l_j \geq n+j-2p+5$, and $j = m$ yields $2n - 6 \geq 2l_m \geq n + m - 2p + 5$; that is, $m \leq n + 2p - 11$. The result for $p = 5$ follows from $2l_j \geq n + j - 6$, see Theorem 3.2. $\square$

7.3.3 Depths of twigs

We show that the skeleton groups at depth $e < \mathfrak{e}_n$ in the body $\mathcal{T}_n$ are exactly the capable groups at depth e in $\mathcal{T}_n$. Recall the definition of $\iota_{n,e}\colon T_n/T_{n+e}\to A_e$, see Definition 6.2.

7.22 Lemma. *If $x_f \in \widehat{\mathrm{L}}(n,e)$ and $e \leq \mathfrak{e}_n$, then*

$$E(\mathfrak{m}_{n,e} + x_f) \cong C_{\tilde{f},e}$$

where $\tilde{f}\colon T\wedge T\to T_n/T_{n+e}$ is defined as $\iota_{n,e}^{-1}\circ f$.

Proof. By construction, $E(\mathfrak{m}_{n,e} + x_f) \cong E(\gamma'_{n,e} + \Gamma_f)$ where $\gamma'_{n,e}$ and Γ_f are the 2-cocycles defined in Sections 7.1.1 and 6.5.3, respectively. By definition, if $u, v \in S_n$, then

$$\gamma'_{n,e}(u,v) = \iota_{n,e}\left(\tau(uv)^{-1}\tau(u)\tau(v)\right)$$

where $\tau\colon S_n \to S_{n+e}$ is the canonical transversal with $\tau(u) = u$. The subgroup T_n/T_{n+e} of $C_{\tilde{f},e}$ is central in $(T/T_{n+e}, \odot)$ and, thus, the following mapping is an isomorphism

$$E(\gamma'_{n,e} + \Gamma_f) \to C_{\tilde{f},e}, \quad (\mathfrak{g}^i u, a) \mapsto (\mathfrak{g}^i, \tau(u)\iota_{n,e}^{-1}(a)) \quad (u \in T/T_n). \qquad \square$$

7.23 Corollary. *The twigs of a body are subtrees of depth 1.*

Proof. Let G be a capable group at depth $e < \mathfrak{e}_n$ in the body $\mathcal{T}_n$. By Theorem 7.11 and Lemma 7.17, the group G is isomorphic to $E(\mathfrak{m}_{n,e} + x_f)$ for some hom tail vector $x_f \in \widehat{L}(n, e)$, and it follows from Lemma 7.22 that G is a skeleton group. Thus, by Lemma 7.19, the capable groups at depth $e < \mathfrak{e}_n$ in the body $\mathcal{T}_n$ are exactly the skeleton groups at depth e, and all twig groups at depth e are terminal. $\qquad \square$

8 Periodicity of type 1

We now prove the periodicity of type 1 as introduced in Theorem 2.6. Throughout this chapter, let $n \geq p+1$ and $1 \leq e \leq \mathfrak{e}_n$. We write the integer n as $n = \mathfrak{x}d + \mathfrak{i}$ with integers $\mathfrak{x} \geq 1$ and $1 \leq \mathfrak{i} \leq d$.

8.1 Graph isomorphisms

It follows from Section 6.5.3 that a tail vector of S_n in A_e can be considered as a tail vector of S_{n+d} in A_e, and vice versa. We denote this isomorphism of tail vectors by

$$\nu_{n,e} \colon \mathcal{Z}(n, e) \to \mathcal{Z}(n+d, e) \quad \text{with} \quad \nu_{n,e}(x) = x$$

and we prove that its restriction to the group $\mathrm{TH}(n, e) = \mathrm{T}(n, e) \oplus \mathrm{H}(n, e)$ is an $\mathrm{Aut}(S)$-module isomorphism between $\mathrm{TH}(n, e)$ and $\mathrm{TH}(n+d, e)$.

First, we examine the homomorphism $\xi_{n,e} \colon \mathrm{Aut}(S) \to \mathrm{Aut}(A_e)$ of Definition 7.12.

8.1 Lemma. *If $\alpha \in \mathrm{Aut}(S)$, then $\xi_{n,e}(\alpha) = \xi_{n+d,e}(\alpha)$.*

Proof. If $\alpha = \alpha(j, c, t)$ as defined in Theorem 6.12, then

$$\xi_{n,e}(\alpha) \colon A_e \to A_e, \quad u \mapsto \sigma_j(u)^{(1+\mathfrak{g}+\ldots+\mathfrak{g}^{j-1})^{\mathfrak{i}-1}c},$$

that is, $\xi_{n,e}$ depends only on the value of n modulo d. □

8.2 Lemma. *The restriction $\nu_{n,e}|_{\mathrm{TH}(n,e)} \colon \mathrm{TH}(n, e) \to \mathrm{TH}(n+d, e)$ is an $\mathrm{Aut}(S)$-module isomorphism.*

Proof. Let $x \in \mathrm{TH}(n, e)$ and $\alpha \in \mathrm{Aut}(S)$ be arbitrary. By Lemma 7.10, the tail vectors x and x^α both have a trivial mainline component. Let γ_0 and γ_1 be the canonical 2-cocycles defined by x and $\nu_{n,e}(x)$, respectively, see Definition 4.8. By construction, the value of $\gamma_i(u, v)$ with $u, v \in S_{n+\mathfrak{i}d}$ and $i = 0, 1$ corresponds to the tail occurring at the collection of the word uv in $E(x)$ and $E(\nu_{n,e}(x))$, respectively. We write $\alpha_i = \alpha|_{S_{n+\mathfrak{i}d}}$ for $i = 0, 1$ and denote the projection $S_{n+d} \to S_n$ by π. It follows from the definition that $\alpha_0 \circ \pi = \pi \circ \alpha_1$. In the remaining part of the proof we show that

$$(*) \quad \gamma_1 = \gamma_0 \circ (\pi, \pi).$$

Then Lemma 8.1 implies that

$$\xi_{n+d,e}(\alpha)^{-1} \circ \gamma_1 \circ (\alpha_1, \alpha_1) = \xi_{n,e}(\alpha)^{-1} \circ \gamma_0 \circ (\alpha_0, \alpha_0) \circ (\pi, \pi),$$

and it follows from Remark 6.15 that $\nu_{n,e}(x^\alpha) = \nu_{n,e}(x)^\alpha$, which proves the lemma.

Let $\tau\colon S_n \to S_{n+d}$ be the canonical transversal with $\tau(s) = s$. The exponent of A_e divides $q = p^{\mathfrak{f}}$, and every element in T_n/T_{n+e} is a word in $\{\mathfrak{t}_1^q, \ldots, \mathfrak{t}_d^q\}$. Thus, $u \in T/T_{n+d}$ can be written as $u = u't_u$ in S_{n+d} where $u' = \tau(\pi(u))$ and $t_u \in T_n/T_{n+e}$, and there occur no tails at the collection of $u't_u$ in $E(\nu_{n,e}(x))$, that is, $\gamma_1(u', t_u) = 1$. In particular, if $\mathfrak{g}^k a, \mathfrak{g}^l b \in S_{n+d}$ with $0 \leq k, l \leq d$ and $a, b \in T/T_{n+d}$, then it follows from the definition that

$$\gamma_1(\mathfrak{g}^k a, \mathfrak{g}^l b) = \gamma_1(a, \mathfrak{g}^l)\gamma_1(\mathfrak{g}^k, \mathfrak{g}^l)\gamma_1(a^{\mathfrak{g}^l}, b).$$

Clearly, $\pi(a^{\mathfrak{g}^l}) = \pi(a)^{\mathfrak{g}^l}$ and, thus, (∗) holds if and only if $\gamma_1(a, \mathfrak{g}^l) = \gamma_0(\pi(a), \mathfrak{g}^l)$ for all $0 \leq l \leq d$ and $a \in T/T_{n+e}$. It is sufficient to consider $l = 1$. Then, by definition, $\gamma_1(a, \mathfrak{g}) = \gamma_1(a', \mathfrak{g})$, and $\gamma_1(a', \mathfrak{g}) = \gamma_0(\pi(a), \mathfrak{g})$ can be deduced from Remark 6.9 since $a' = \pi(a)$ as words. □

An important consequence is the following corollary.

8.3 Corollary. *A set of* $\mathrm{Aut}(S)$-*orbit representatives in* $\widehat{\mathrm{TH}}(n, e)$ *is mapped under* $\nu_{n,e}$ *onto a set of* $\mathrm{Aut}(S)$-*orbit representatives in* $\widehat{\mathrm{TH}}(n + d, e)$.

Definition. For a group G in the body $\mathcal{T}_n$ with $G \cong E(\mathfrak{m}_{n,e} + y)$ for some $y \in \widehat{\mathrm{TH}}(n, e)$ we define

$$\iota(G) = E(\nu_{n,e}(\mathfrak{m}_{n,e} + y))$$

and, thus, obtain a mapping $\iota\colon (\mathcal{T}_{p+1} \cup \mathcal{T}_{p+2} \cup \ldots) \to \mathcal{T}$. The restriction of ι to the body $\mathcal{T}_n$ is denoted by $\iota_n = \iota|_{\mathcal{T}_n}$.

We show that ι is well-defined. If $y_1, y_2 \in \widehat{\mathrm{TH}}(n, e)$ with $E(\mathfrak{m}_{n,e} + y_1) \cong G \cong E(\mathfrak{m}_{n,e} + y_2)$, then there exists $\alpha \in \mathrm{Aut}(S)$ with $y_1^\alpha = y_2$ and Lemma 8.2 shows that $\nu_{n,e}(y_2) = \nu_{n,e}(y_1)^\alpha$. It follows from Lemma 7.4 that $\nu_{n,e}(\mathfrak{m}_{n,e}) = \mathfrak{m}_{n+d,e}$ and, hence,

$$E(\mathfrak{m}_{n+d,e} + \nu_{n,e}(y_1)) \cong \iota(G) \cong E(\mathfrak{m}_{n+d,e} + \nu_{n,e}(y_2)).$$

Recall that $\mathcal{T}_{n+d}[\mathfrak{e}_n]$ is the shaved subtree of the body $\mathcal{T}_{n+d}$ induced by the groups in $\mathcal{T}_{n+d}$ of distance at most $\mathfrak{e}_n$ from S_{n+d}. The following theorem describes the periodicity of type 1 as introduced in Theorem 2.6.

8.4 Theorem. *The mapping*

$$\iota_n\colon \mathcal{T}_n \to \mathcal{T}_{n+d}$$

is an embedding of rooted trees with $\iota_n(\mathcal{T}_n) = \mathcal{T}_{n+d}[\mathfrak{e}_n]$.

Proof. It follows from Corollaries 7.15 and 8.3 that ι_n is a bijection between the groups at depth e in $\mathcal{T}_n$ and the groups at depth e in $\mathcal{T}_{n+d}$ for all $e \leq \mathfrak{e}_n$. Thus, it remains to show that ι_n is a homomorphism of rooted trees. Clearly, ι_n maps S_n onto S_{n+d}, and all groups at depth 1 in $\mathcal{T}_n$ and $\mathcal{T}_{n+d}$ are immediate descendants of S_n and S_{n+d}, respectively. For $e \geq 2$ we consider the projection $\pi\colon A_e \to A_{e-1}$. By Lemmas 7.2 and 7.4, the parent of $E(\mathfrak{m}_{n,e} + y)$ is $E(\mathfrak{m}_{n,e-1} + \pi(y))$, and the theorem follows from $\pi \circ \nu_{n,e} = \nu_{n,e-1} \circ \pi$. □

This periodicity is sketched in Figure 8.1. Recall the definition of $\mathfrak{c}$, see Definition 2.1.

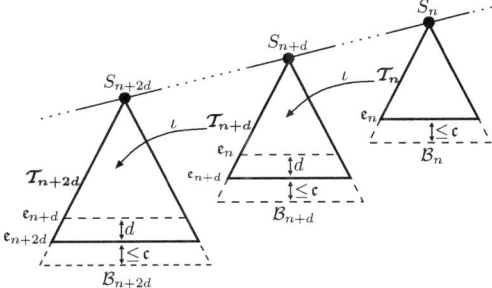

Figure 8.1: The periodicity of type 1.

8.2 Periodicity classes

The periodicity described in Theorem 8.4 is a graph theoretic periodicity. However, this graph theoretic pattern is reflected in the structure of the groups involved. In this section, we define the periodicity class of a group and show that all groups in this class can be described by a single parameterized presentation with one integer parameter.

8.5 Definition. The periodicity class of a group G in the body $\mathcal{T}_n$ is the infinite sequence of groups
$$\mathcal{P}(G) = (G, \iota(G), \iota^2(G), \ldots).$$

8.6 Theorem. *The groups in a periodicity class can be described by a single parameterized presentation with one integer parameter.*

Proof. Let G be a group in the body $\mathcal{T}_n$. We can assume that $G = E(x)$ with $x = \mathfrak{m}_{n,e} + y$ for some $y \in \widehat{\mathrm{TH}}(n,e)$. Since $\nu_{n,e}(y) = y$, we regard y as a tail vector of S_{n+kd} in $\mathcal{A}_e$ for all $k \geq 0$. Thus, by definition, the periodicity class of G is $\mathcal{P}(G) = (G_k \mid k \geq 0)$ where $G_k = E(\mathfrak{m}_{n+kd,e} + y)$ for all $k \geq 0$. Recall that $\mathfrak{m}_{n,e} = \mathfrak{m}_{n+kd,e}$ for all $k \geq 0$. Hence, if $x = (x_{i,j})_{0 \leq i \leq j \leq d}$, then Definition 6.14 shows that G_k is the group defined by the consistent p.c.p. with generators $\mathcal{S} \cup \mathcal{A}_e$ and relations

$$\{\mathfrak{t}_u^{\mathfrak{t}_l} = \mathfrak{t}_u x_{l,u}, \ \mathfrak{t}_j^{\mathfrak{g}} = \mathfrak{t}_{j+1} x_{0,j}, \ \mathfrak{t}_d^{\mathfrak{g}} = \mathfrak{t}_1^{l_{n+kd,1}} \ldots \mathfrak{t}_d^{l_{n+kd,d}} x_{0,d} \mid l < u \text{ and } j < d\} \cup$$
$$\{\mathfrak{g}^p = x_{0,0}, \ \mathfrak{t}_j^{p^{\mathfrak{x}+k}} = w_{n+kd,j} x_{j,j}, \ \mathfrak{t}_l^{p^{\mathfrak{x}+k+1}} = x_{l,l} \mid j \leq d - \mathfrak{i} + 1 < l\} \cup \mathcal{C}_e \cup \mathcal{M}_e.$$

By Remark 6.9, the exponents of these relations are arithmetic expressions containing the integer k as parameter. This proves the theorem. □

9 Periodicity of type 2

It is shown in Chapter 8 that the body $\mathcal{T}_n$ with $n \geq p+1$ can be embedded into $\mathcal{T}_{n+d}$ such that $\mathcal{T}_n$ and the shaved subtree $\mathcal{T}_{n+d}[\mathfrak{e}_n]$ are isomorphic as rooted trees. In order to describe $\mathcal{T}_{n+d}$ completely, it remains to describe its growth, that is, the d levels of groups at depth $\mathfrak{e}_n + 1, \ldots, \mathfrak{e}_n + d$ in $\mathcal{T}_{n+d}$. Since the widths of the bodies are unbounded in general, we cannot expect this subgraph of $\mathcal{T}_{n+d}$ to be isomorphic to a subgraph of $\mathcal{T}_n$. Motivated by a conjecture made in [15], we conjecture that this graph can be described by another periodic pattern. Recall that the d-step descendant tree $\mathcal{D}_d(G)$ of a group G in $\mathcal{T}$ is the subtree of $\mathcal{T}$ generated by the descendants of distance at most d from G.

9.1 Conjecture. *There is an integer $n_0 = n_0(p)$ with the following property: If $n \geq n_0$ and G is a group at depth $\mathfrak{e}_n$ in $\mathcal{T}_{n+d}$, then there exists a group H at depth $\mathfrak{e}_{n-d}$ in $\mathcal{T}_{n+d}$ such that*

$$\mathcal{D}_d(G) \cong \mathcal{D}_d(H)$$

as rooted trees. The group H is called a periodic parent of G.

The periodicity described in Conjecture 9.1 is called the periodicity of type 2 and it is depicted in Figure 9.1. If this conjecture holds, then there might exist a *natural* mapping which chooses a periodic parent of a given group. This would allow us to describe the bodies of the tree $\mathcal{T}$ by a finite subgraph and periodic patterns, which supports Conjecture 1.2.

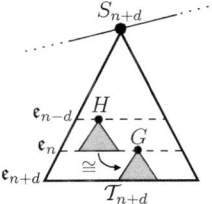

Figure 9.1: A periodic parent of G.

In order to confirm Conjecture 9.1, one has to construct and compare descendant trees. Hence, in the first part of this chapter, we discuss this construction in more detail. At the end of this chapter, we prove Conjecture 9.1 for some significant special cases.

Clearly, if the group G is terminal, then its descendant tree $\mathcal{D}_d(G)$ consists only of the vertex G. Therefore, we consider only descendant trees of capable groups in the following.

9.1 Descendant trees

Let $m = n+d$ with $n \geq p+1$, and let $e \geq 1$ and $k \geq 0$ be integers such that $e+k \leq \mathfrak{e}_m$. We use this notation and describe the construction of the k-step descendant tree of a capable group G at depth e in $\mathcal{T}_m$. By Theorem 7.11 and Lemma 7.17, we can assume that $G = E(\mathfrak{m}_{m,e} + x_f)$ for some hom tail vector x_f defined by a surjective and liftable P-homomorphism f from $T \wedge T$ to A_e. We identify f with a lifting $T \wedge T \to T$.

First, we introduce some more notation. Recall that an automorphism $\alpha \in \operatorname{Aut}(S)$ acts on $\operatorname{TH}(m,e)$ via the compatible pair $(\alpha, \xi_{m,e}(\alpha)) \in \Sigma_{m,e}$, see Remark 7.14.

Definition. Let $f\colon T \wedge T \to T$ be a P-homomorphism.

a) Let $x_{f,e} \in \operatorname{L}(m,e)$ be defined as $x_{f,e} = x_{\pi \circ f}$ where $\pi\colon T \to A_e$ is the projection.

b) Let $E_{m,e}(f) = E(\mathfrak{m}_{m,e} + x_{f,e})$ be the extension defined by $\mathfrak{m}_{m,e} + x_{f,e}$.

c) Let $\Sigma_{m,e}(f) = \operatorname{Stab}_{\operatorname{Aut}(S)}(x_{f,e})$ be the stabilizer of $x_{f,e}$.

A group H in $\mathcal{T}_m$ is a **k-step descendant** of G if there is a path of length k from G to H. The k-step descendants of G are determined up to isomorphism in the following lemma.

Lemma. *Let $\pi\colon A_{e+k} \to A_e$ be the projection.*

a) *A complete set of isomorphism types of k-step descendants of $E_{m,e}(f)$ is given by*

$$\{E(\mathfrak{m}_{m,e+k} + y) \mid y \in \widehat{\operatorname{TH}}(m,e+k) \text{ with } \pi(y) = x_{f,e}\}.$$

b) *Two groups $E(\mathfrak{m}_{m,e+k}+y_1)$ and $E(\mathfrak{m}_{m,e+k}+y_2)$ as in part a) are isomorphic if and only if $y_1^\alpha = y_2$ for some $\alpha \in \Sigma_{m,e}(f)$.*

Proof. a) Let H be a k-step descendant of G. By Theorem 7.11, we can assume that $H = E(\mathfrak{m}_{m,e+k} + z)$ for some tail vector $z \in \widehat{\operatorname{TH}}(m,e+k)$ and it follows from Lemmas 7.2 and 7.4 and Theorem 7.11 that $\pi(z)^\alpha = x_{f,e}$ for some automorphism $\alpha \in \operatorname{Aut}(S)$. If $y = z^\alpha$, then H and $E(\mathfrak{m}_{m,e+k} + y)$ are isomorphic and $\pi(y) = \pi(z)^\alpha = x_{f,e}$ follows from $\pi \circ \gamma_z^\alpha = (\pi \circ \gamma_z)^\alpha$. Now Lemma 7.2 proves the assertion.

b) This follows from Theorem 7.11. $\square$

We analyze this lemma in more detail and, therefore, we need some more notation. Recall that the group of hom tail vectors can be written as $\operatorname{H}(m,e) = \operatorname{L}(m,e) \oplus \operatorname{H}_t(m,e)$ where $\operatorname{H}_t(m,e)$ is cyclic of order p, see Corollary 6.30.

Definition. Let $\mathcal{Z}(m,e,k) = \mathcal{Z}_l(m,e,k) \oplus \mathcal{Z}_t(m,e,k)$ where

$$\mathcal{Z}_l(m,e,k) = \{x_{h,e+k} \in \operatorname{L}(m,e+k) \mid h \in \operatorname{Hom}_P(T \wedge T, T_{e+1})\} \quad \text{and}$$

$$\mathcal{Z}_t(m,e,k) = \operatorname{T}(m,e+k) \oplus \operatorname{H}_t(m,e+k).$$

By definition, every $y \in \widehat{\operatorname{TH}}(m,e+k)$ with $\pi(y) = x_{f,e}$ can be written as $y = x_{f,e+k} + z$ for some $z \in \mathcal{Z}(m,e,k)$. As shown above, two tail vectors $\mathfrak{m}_{m,e+k} + x_{f,e+k} + z_1$ and $\mathfrak{m}_{m,e+k} + x_{f,e+k} + z_2$ with $z_1, z_2 \in \mathcal{Z}(m,e,k)$ define isomorphic extensions if and only if there exists $\alpha \in \Sigma_{m,e}(f)$ with

$$z_2 = z_1^\alpha + x_{f,e+k}^\alpha - x_{f,e+k}.$$

The following definition takes this into account.

9.2. The action of the stabilizer

Definition. We say that $\alpha \in \Sigma_{m,e}(f)$ acts on $y \in \mathcal{Z}(m,e,k)$ via
$$y^{\tilde{\alpha}} = y^{\alpha} + x_{f,e+k}^{\alpha-1},$$
and we call y^{α} and $x_{f,e+k}^{\alpha-1}$ the linear and affine part of this action, respectively.

Clearly, this "action" is not a group action since $y \mapsto y^{\tilde{\alpha}}$ is not a homomorphism. However, if $y \in \mathcal{Z}(m,e,k)$ and $\alpha, \beta \in \Sigma_{m,f}(f)$, then $y^{\widetilde{\alpha \circ \beta}} = (y^{\tilde{\alpha}})^{\tilde{\beta}}$ and $y^{\tilde{\mathrm{id}}} = y$.

The main result of this section is summarized in the following corollary.

9.2 Corollary. *Let $\mathcal{M}$ be a set of $\Sigma_{m,e}(f)$-orbit representatives in $\mathcal{Z}(m,e,k)$. Then, up to isomorphism, the k-step descendants of $E_{m,e}(f)$ are*
$$(E(\mathfrak{m}_{m,e+k} + x_{f,e+k} + y) \mid y \in \mathcal{M}).$$

9.2 The action of the stabilizer

Again, we consider a surjective P-homomorphism $f: T \wedge T \to T$ and write $m = n + d$ with $n \geq p + 1$. Let $e \geq 1$ and $k \geq 0$ be integers such that $e + k \leq \mathfrak{e}_m$. With regard to Corollary 9.2, we now analyze the action of the stabilizer $\Sigma_{m,e}(f)$ on the group $\mathcal{Z}(m,e,k)$.

We write $m = \mathfrak{r}d + \mathfrak{i}$ with integers $\mathfrak{r} \geq 1$ and $1 \leq \mathfrak{i} \leq d$, and we denote the projections from T and A_{e+k} to A_e, respectively, by π for all k.

9.2.1 The action on $\mathcal{Z}_l(m,e,k)$

The following corollary is a consequence of Lemmas 7.10 and 7.16.

Corollary. *The group $\mathcal{Z}_l(m,e,k)$ is invariant under the action of $\Sigma_{m,e}(f)$.*

Proof. Let $y = x_{h,e+k}$ for some P-homomorphism $h: T \wedge T \to T_{e+1}$ and let $\alpha \in \Sigma_{m,e}(f)$. It follows from the definition and Lemma 7.16 that $\pi(y^{\tilde{\alpha}}) = 0$ and $y^{\tilde{\alpha}} \in \mathrm{L}(m, e+k)$. □

Lemma 7.16 and Remark 7.14 imply that $\Sigma_{m,e}(f) = \mathrm{Stab}_{\mathrm{Aut}(S)}(\pi \circ f)$ where $\alpha \in \mathrm{Aut}(S)$ acts on $\pi \circ f$ via
$$(\pi \circ f)^{\alpha} = \xi_{m,e}(\alpha)^{-1} \circ \pi \circ f \circ (\alpha \wedge \alpha)|_{T \wedge T}.$$

We write
$$H_e = \mathrm{Hom}_P(T \wedge T, T_{e+1})$$
and identify H_0/H_e with the group of liftable P-homomorphisms $T \wedge T \to A_e$. Then
$$\Sigma_{m,e}(f) = \mathrm{Stab}_{\mathrm{Aut}(S)}(f + H_e)$$
where $\alpha = \alpha(j,c,t) \in \mathrm{Aut}(S)$ with $l = j^{-1} \bmod p$ acts on $f + H_e$ as
$$(f + H_e)^{\alpha}: u \wedge v \mapsto \sigma_l(f(\alpha(u) \wedge \alpha(v)))^{(1+\theta+\ldots+\theta^{l-1})^{i-1}\sigma_l(c)^{-1}} T_{e+1}.$$

Hence, if $j = 1$, then $(f + H_e)^{\alpha} = f^{\alpha} + H_e$ where $f^{\alpha} = \alpha^{-1} \circ f \circ (\alpha \wedge \alpha)|_{T \wedge T}$.

Using this notation, we can write
$$x_{h,e+k}^{\alpha} = x_{(h+H_{e+k})^{\alpha}}$$
for $\alpha \in \mathrm{Aut}(S)$ and $x_{h,e+k} \in \mathrm{L}(m, e+k)$.

The following remark explains the advantage of writing $h: T \wedge T \to T$ as a $\mathbb{Q}_p(\theta)$-linear combination of the homomorphisms $F_2, \ldots, F_{d/2}$, see Lemma 5.10.

9.3 Remark. Let $2 \leq a \leq d/2$ and $\alpha = \alpha(j, z, t)$ with $l = j^{-1} \bmod p$. Recall that $m = \mathfrak{r}d + \mathfrak{i}$, and a straightforward computation shows that

$$(cF_a + H_e)^\alpha = (cF_a)^\alpha + H_e$$

where

$$(cF_a)^\alpha = (1 + \theta + \ldots + \theta^{l-1})^{\mathfrak{i}-1}\sigma_l(\rho_a(z))\sigma_l(c)F_a$$

with $\rho_a(z) = z^{-1}\sigma_a(z)\sigma_{1-a}(z)$ as in Section 5.4, cf. Section A.1.2. Using this action, we can define h^α for every P-homomorphism $h: T \wedge T \to T$ written as $h = c_2F_2 + \ldots + c_{d/2}F_{d/2}$, which allows us to define $(h + H_e)^\alpha = h^\alpha + H_e$, that is, $x_{h,e+k}^\alpha = x_{(h^\alpha + H_{e+k})}$.

Note that the action of $\alpha(1, z, t) \in \mathrm{Aut}(S)$ coincides with the action of the unit $z \in \mathcal{U}_p$ as defined in Section 5.4, which motivates the investigation of the stabilizer $\mathrm{Stab}_{\mathcal{U}_p^{(2)}}(f + H_e)$ in Theorem 5.14.

We conclude this paragraph with a corollary to Remark 9.3.

9.4 Corollary. *If $p \geq 7$, then the bodies of the coclass tree $\mathcal{T}(p)$ have unbounded width.*

Proof. If G is a capable group at depth $1 \leq e \leq \mathfrak{e}_m - 2$ in the body $\mathcal{T}_m$, then $G \cong E_{m,e}(f)$ for some surjective P-homomorphism $f: T \wedge T \to T$, and $E_{m,\mathfrak{e}_m}(f)$ is a descendant of G at depth $\mathfrak{e}_m$. We now consider a surjective P-homomorphism $h = \sum_{a=2}^{d/2} c_a F_a$ with $c_{a'} = 0$ for some $2 \leq a' \leq d/2$ defining a capable group $H = E_{m,e}(h)$ at depth e in $\mathcal{T}_m$. By Corollary 9.2, the capable immediate descendants of H correspond to the $\Sigma_{m,e}(h)$-orbits on $\mathcal{Z}_l(m, e, 1)$ and, for a contradiction, we assume that H has only one immediate descendant, that is, $\mathcal{Z}_l(m, e, 1) = \{x_{h,e+1}^{\alpha-1} \mid \alpha \in \Sigma_{m,e}(h)\}$. Hence, if $j \geq 0$, then every homomorphism $f: T \wedge T \to T_{e+1+j}$ can be written as $f = \kappa^j(h^\alpha - h) + f'$ for some $\alpha \in \Sigma_{m,e}(h)$ and $f': T \wedge T \to T_{e+2+j}$, where h^α is defined as in Remark 9.3. This shows that we can write $\kappa^{e-1}F_{a'} = \sum_{j \geq 0} \kappa^j(h^{\alpha_j} - h)$ for certain $\alpha_0, \alpha_1, \ldots$ lying in $\Sigma_{m,e}(h)$. Thus, $\kappa^{e-1}F_{a'}$ is a $\mathbb{Q}_p(\theta)$-linear combination of $\{F_a \mid 2 \leq a \leq d/2, \, a \neq a'\}$, contradicting Lemma 5.10. $\square$

9.2.2 The action on $\mathcal{Z}_t(m, e, k)$

The group $\mathcal{Z}_t(m, e, k)$ is the direct sum of the group of twig tail vectors $\mathrm{T}(m, e + k)$ and a summand $\mathrm{H}_t(m, e + k)$ of order p. It follows from Remark 7.14 that $(x_h)^\alpha - x_{h^\alpha}$ is a twig tail vector whenever $x_h \in \mathrm{H}_t(m, e + k)$ and $\alpha \in \mathrm{Aut}(S)$. Since $x_{h^\alpha} \notin \mathrm{H}_t(m, e + k)$ in general, the group $\mathcal{Z}_t(m, e, k)$ is not $\mathrm{Aut}(S)$-invariant.

We now describe the linear action of $\mathrm{Aut}(S)$ on the group of twig tail vectors. For this purpose, we consider $\mathrm{T}(m, e + k) \cong \mathbb{F}_p^2$ as an $\mathbb{F}_p$-vector space with basis consisting of two tail vectors such that one has non-trivial tail $x_{0,0}$ and the other one has non-trivial tail $x_{0,d}$ only. Recall that $m = \mathfrak{r}d + \mathfrak{i}$ with $1 \leq \mathfrak{i} \leq d$.

9.5 Lemma. *The automorphism $\alpha(l, \theta^a\omega^b c, t)$ with $c \in \mathcal{U}_p^{(2)}$ acts on $\mathrm{T}(m, e + k) \cong \mathbb{F}_p^2$ as*

$$(\omega^{-b}l^{3-e-k-\mathfrak{i}} \bmod p) \begin{pmatrix} 1 & 0 \\ t \bmod (\mathfrak{p}, +) & \omega^b l^{-1} \bmod p \end{pmatrix} \in \mathrm{GL}(2, \mathbb{F}_p).$$

9.2. The action of the stabilizer

Proof. We write $\alpha = \alpha(l, z, t)$ with $z = \theta^a \omega^b c$. Let $x \in T(m, e+k)$ be a twig tail vector with canonical 2-cocycle $\gamma = \gamma_x$ and write $y = x^\alpha$ for the tail vector defined by γ^α, see Remark 6.15. By construction, the value of $\gamma(u,v)$ with $u, v \in S_m$ corresponds to the tail occurring at the collection of the word uv in $E(x)$. This shows that y has non-trivial tails $y_{0,0}, y_{0,1}, \ldots, y_{0,d}$ only. Note that $\xi_{m,e+k}(\alpha)^{-1}(u) = u^{j^{e+k+i-2}\omega^{-b}}$ for all $u \in T_{e+k}/T_{e+k+1}$ where $j = l^{-1} \bmod p$.

If x has a non-trivial tail $x_{0,0} = u$ only, then $y_{0,1} = \ldots = y_{0,d} = 1$, and it follows from $|\{1 \leq s \leq d \mid (ls \bmod p) + l \geq p\}| = l$ that

$$y_{0,0} = \xi_{m,e+k}(\alpha)^{-1}(u^l) = u^{j^{e+k+i-3}\omega^{-b}}.$$

Now let x be a twig tail vector with non-trivial tail $x_{0,d} = u$ only. We write $\gamma^\alpha = \gamma^{\alpha(1,1,t) \circ \alpha(1,z,1) \circ \alpha(l,1,1)}$ and make a case distinction:

First, we consider $\alpha = \alpha(1,1,t)$ with $t = \mathbf{t}_1^{a_1} \ldots \mathbf{t}_d^{a_d}$ for some $a_1, \ldots, a_d \in \mathbb{Z}_p$. By induction, the exponent of $\mathbf{t}_d$ in t^{θ^s} with $0 < s < d$ is congruent to $a_{d-s} - a_{p-s}$ modulo p and, thus,

$$y_{0,0} = \prod_{s=1}^{d} \gamma(\mathbf{g}^s t^{1+\theta+\ldots+\theta^{s-1}}, \mathbf{g}) = u^{(a_1 + \ldots + a_d)} = u^{t \bmod (\mathfrak{p}, +)}.$$

It follows from $y_{0,s} = \gamma(\mathbf{t}_s, gt)$ that $y_{0,1} = \ldots = y_{0,d-1} = 1$ and $y_{0,d} = u$.

If $\alpha = \alpha(1,z,1)$, then $y_{0,0} = 1$ and $C_p(x)$ is invariant under α. Thus, if $z = \theta^a c$, then α acts trivially and $y = x^\alpha = x$. If $z = \omega^b$, then $y = x^\alpha = x$ as well.

Finally, we consider $\alpha = \alpha(l,1,1)$ with $2 \leq l \leq d$. Clearly, the tail $y_{0,0}$ induced by γ^α is trivial. If $1 \leq s \leq d$ and $j = l^{-1} \bmod p$, then $y_{0,s} = \gamma(\sigma_l(\mathbf{t}_s), \mathbf{g}^l)^{j^{e+k+i-2}}$ and

$$\gamma(\sigma_l(\mathbf{t}_s), \mathbf{g}^l) = \begin{cases} u & : s \equiv -l^{-1} \bmod p, \\ u^{-1} & : s \equiv -l^{-1} + 1 \bmod p, \\ 1 & : \text{otherwise}. \end{cases}$$

Hence, modulo coboundary tails, $y_{0,1} = \ldots = y_{0,d-1} = 1$ and

$$y_{0,d} = \left(\gamma(\sigma_l(\mathbf{t}_d), \mathbf{g}^l) \prod_{s=1}^{d-1} \gamma(\sigma_l(\mathbf{t}_s), \mathbf{g}^l)^{-s}\right)^{j^{e+k+i-2}},$$

see Lemma 6.22, which implies that $y_{0,d} = u^{j^{e+k+i-2}}$. □

9.2.3 An $\text{Aut}(S)$-module isomorphism

Again, we consider the linear action on $\mathcal{Z}(m, e, k)$. Let $1 \leq j \leq \mathfrak{e}_m - d - k$ be an integer.

Definition. The isomorphism $T_{j+1}/T_{j+k+1} \to T_{j+d+1}/T_{j+k+d+1}$ defined by multiplication by p induces an isomorphism

$$\phi_{m,j,k} \colon \mathcal{Z}(m, j, k) \to \mathcal{Z}(m, j+d, k).$$

Its restrictions to $\mathcal{Z}_l(m, j, k)$ and $\mathcal{Z}_t(m, j, k)$ are denoted by $\phi_{m,j,k}^l$ and $\phi_{m,j,k}^t$, respectively.

We now show that $\phi_{m,j,k}$ is compatible with the $\mathrm{Aut}(S)$-action.

9.6 Lemma. *If $\alpha \in \mathrm{Aut}(S)$ and $x \in \mathcal{Z}(m,j,k)$, then $\phi_{m,j,k}(x^\alpha) = \phi_{m,j,k}(x)^\alpha$.*

Proof. First, we show that $y = x^\alpha$ lies in $\mathcal{Z}(m,j,k)$ and, by Lemmas 7.16 and 9.5, it remains to consider $x_f \in \mathrm{H}_t(m,j+k)$. Then $x_f^\alpha \in \mathcal{Z}(m,j,k)$ follows from Remark 7.14 since $x_f^\alpha - x_{f^\alpha}$ is a twig tail vector. Now let $x \in \mathcal{Z}(m,j,k)$ and denote by ϕ the isomorphism from T_{j+1}/T_{j+k+1} to $T_{j+d+1}/T_{j+k+d+1}$ induced by multiplication by p. Then the assertion follows from $\phi \circ (\gamma_x^\alpha) = (\phi \circ \gamma_x)^\alpha$ where γ_x is the canonical 2-cocycle defined by x. □

9.7 Lemma. *If $N = \{\alpha(1,z,t) \mid z \in C_p(\theta) \times \mathcal{U}_p^{(d)}, t \in T\}$, then $\phi_{m,j,k}^t$ is an N-module isomorphism and $\alpha(1,z,1) \in N$ acts trivially on $\mathcal{Z}_t(m,j,k)$.*

Proof. It follows from Lemma 9.5 that $\phi_{m,j,k}|_{\mathrm{T}(m,j+k)}$ is an N-module isomorphism. If $x_h \in \mathrm{H}_t(m,j+k)$ and $\alpha \in N$, then Remark 7.14 shows that $(x_h)^\alpha - x_{h^\alpha}$ is a twig tail vector, and $x_{h^\alpha} = x_h$ since $z \equiv 1 \bmod p$. Hence, $\mathcal{Z}_t(m,j,k)$ is N-invariant, and it is easy to see that $\alpha(1,z,1) \in N$ acts trivially. Now Lemma 9.6 proves the assertion. □

The following remark summarizes our approach to attack Conjecture 9.1.

Remark. Let $G = E_{m,\mathfrak{e}_n}(f)$ and $H = E_{m,\mathfrak{e}_{n-d}}(h)$ be capable groups at depth $\mathfrak{e}_n$ and $\mathfrak{e}_{n-d}$, respectively, in the body $\mathcal{T}_m$. It follows from the construction that H is a periodic parent of G if the isomorphism $\phi_{m,\mathfrak{e}_{n-d},k}$ is compatible with the action of the stabilizers $\Sigma_{m,\mathfrak{e}_{n-d}}(h)$ and $\Sigma_{m,\mathfrak{e}_n}(f)$, respectively, for all $1 \leq k \leq d$. Thus, given G, the problem is to determine a homomorphism h with this property. The *favored candidate* for h is probably $h = f$ because in this case H would be the d-step parent of G. We examine this problem in the following section.

9.3 The case $p \equiv 5 \bmod 6$

We now prove Conjecture 9.1 in a special case. For this purpose, we start with a preparing paragraph. Again, let $m = n + d$ with $n \geq p + 1$.

9.3.1 Galois complements of a stabilizer

Let $p \geq 5$ be a prime and let $1 \leq e \leq \mathfrak{e}_m$ be an integer. Recall that $p^\star = -(p-3)^2/4$. For an automorphism $\alpha \in \mathrm{Aut}(S_{m+e})$ we define $\xi_{m,e}(\alpha) \in \mathrm{Aut}(A_e)$ similarly as in Definition 7.12. Thus, analogously to an element of $\mathrm{Aut}(S)$, the automorphism α acts on tail vectors via the compatible pair $(\alpha|_{S_m}, \xi_{m,e}(\alpha))$.

Definition. If $f\colon T \wedge T \to T$ is a P-homomorphism, then the projection of $\Sigma_{m,e}(f)$ into $\mathrm{Aut}(S_{m+e})$ is denoted by
$$\widehat{\Sigma}_{m,e}(f) = \mathrm{Stab}_{\mathrm{Aut}(S_{m+e})}(x_{f,e}).$$

9.8 Lemma. *Let $f\colon T \wedge T \to T$ be a surjective P-homomorphism.*
a) *The p'-parts of $|\mathrm{Aut}(E_{m,e}(f))|$ and $|\widehat{\Sigma}_{m,e}(f)|$ are the same.*
b) *If $e \geq -p^\star$, then the kernel of the projection from $\widehat{\Sigma}_{m,e}(f)$ to $\mathrm{Aut}(P)$ is a p-group.*

9.3. The case $p \equiv 5 \bmod 6$

Proof. a) By Lemma 4.4, the homomorphism $\mathrm{Aut}(E_{m,e}(f)) \to \mathrm{Aut}(S_m) \times \mathrm{Aut}(A_e)$ has image $\mathrm{Stab}_{\mathrm{Comp}(S_m, A_e)}(\mathfrak{m}_{m,e} + x_{f,e})$ and its kernel is the p-group $Z^1(S_m, A_e)$, cf. Lemma 6.18. Now the assertion follows from

$$\mathrm{Stab}_{\mathrm{Comp}(S_m, A_e)}(\mathfrak{m}_{m,e} + x_{f,e}) \cong \widehat{\Sigma}_{m,e}(f),$$

see Lemma 7.13 and Remark 7.14.

b) Let N be the kernel of the projection from $\widehat{\Sigma}_{m,e}(f)$ to $\mathrm{Aut}(P)$. If $\alpha \in N$, then $\alpha = \alpha(1, c, t)|_{S_{m+e}}$ for some unit $c \in \mathcal{U}_p$ and $t \in T$. Hence, it remains to show that $c \in C_p(\theta) \times \mathcal{U}_p^{(2)}$, cf. Section 5.1.3. We write $f = \sum_{a=2}^{d/2} c_a F_a$ and $c = \theta^u \omega^v s$ with $s \in \mathcal{U}_p^{(2)}$, and we define $s_a \in \mathfrak{p}^2$ by $\rho_a(s) = 1 + s_a$ for $2 \leq a \leq d/2$, see Section 5.4 . Using this notation, if follows from $\alpha \in \widehat{\Sigma}_{m,e}(f)$ that $\sum_{a=2}^{d/2} c_a (\omega^v + \omega^v s_a - 1) F_a \in \mathrm{Hom}_P(T \wedge T, T_{e+1})$. The element $\omega^v + \omega^v s_a - 1$ is a unit unless $\omega^v = 1$ and, as f is surjective, it follows from Lemma 5.10 that $v \equiv 0 \bmod d$ if $e \geq -p^\star$. This proves the lemma. □

Definition. Let $f \colon T \wedge T \to T$ be a surjective P-homomorphism and $e \geq -p^\star$. By Lemma 9.8, the kernel of the projection from $\widehat{\Sigma}_{m,e}(f)$ to $\mathrm{Aut}(P)$ is a p-group and, thus, has a complement by the Schur-Zassenhaus Theorem. Note that $\alpha(1, 1, t)|_{S_{m+e}} \in \widehat{\Sigma}_{m,e}(f)$ for all $t \in T$, and, hence, this complement can be chosen such that a generator is of the type $\alpha(j, c, 1)|_{S_{m+e}}$. We call the subgroup of $\mathrm{Aut}(S)$ generated by $\alpha(j, c, 1)$ a Galois complement of $\Sigma_{m,e}(f)$.

Thus, if $e \geq -p^\star$, then every $\beta \in \Sigma_{m,e}(f)$ can be written as $\beta = \alpha^i \varepsilon$ where α is a generator of a Galois complement of $\Sigma_{m,e}(f)$, $0 \leq i \leq d$, and ε is of the type $\varepsilon = \alpha(1, z, t)$ with $z \in C_p(\theta) \times \mathcal{U}_p^{(2)}$. The following corollary is a consequence of Lemma 9.8.

9.9 Corollary. *Let G be a group at depth $e \geq -p^\star$ in $\mathcal{T}_m$. If $\mathrm{Aut}(G)$ is a p-group and H is a capable descendant of G in $\mathcal{T}_m$, then $\mathrm{Aut}(H)$ is a p-group.*

Proof. We can assume that $G \cong E_{m,e}(f)$ and $H \cong E_{m,e+i}(f)$ for some surjective P-homomorphism $f \colon T \wedge T \to T$ and $i \geq 0$. If $\alpha \in \mathrm{Aut}(S)$ is a generator of a Galois complement of $\Sigma_{m,e+i}(f)$, then $\alpha|_{S_m}$ lies in $\widehat{\Sigma}_{m,e}(f)$. Now the assertion follows from Lemma 9.8. □

9.3.2 Periodic parents in the case $p \equiv 5 \bmod 6$

The aim of this paragraph is to prove Conjecture 9.1 in the special case described in the following theorem, cf. Figure 9.2. Our proof relies on Theorem 5.14, which is the reason why we restrict attention to a prime $p \equiv 5 \bmod 6$. We use the notation introduced in Section 9.2; recall that $H_e = \mathrm{Hom}_P(T \wedge T, T_{e+1})$.

9.10 Theorem. *Let $p \equiv 5 \bmod 6$ and let G be a group at depth $\mathfrak{e}_n$ in the body $\mathcal{T}_{n+d}$. There is a positive integer $n_0 = n_0(p)$ such that for all $n \geq n_0$ the following holds. If there is a maximal path $K_0 \to K_1 \to \ldots$ in $\mathcal{T}_{n+d}$ such that G has distance $k \leq d$ from a group on this path and if the p'-part of the order of $\mathrm{Aut}(K_s)$ is the same for all $\mathfrak{e}_{n-d} - k \leq s \leq \mathfrak{e}_{n+d}$, then G has a periodic parent $\Pi(G)$ of distance k from this path.*

Proof. Let $m = n + d$ and $e = \mathfrak{e}_n$. We can assume that the groups K_i with $0 \leq i \leq \mathfrak{e}_m$ are defined by a surjective P-homomorphism $f \colon T \wedge T \to T$; that is, $K_i = E_{m,i}(f)$ for all i. Thus, by Corollary 9.2, we can assume that $G = E(\mathfrak{m}_{m,e} + x_{f,e} + y)$ for some $y \in \mathcal{Z}(m, e-k, k)$.

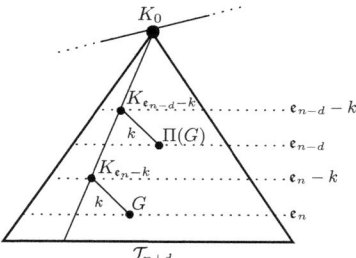

Figure 9.2: The group G and a periodic parent $\Pi(G)$.

Let $y' \in \mathcal{Z}(m, e-k-d, k)$ be a preimage of y under the isomorphism $\phi_{m,e-k-d,k}$ and define $H = E(\mathfrak{m}_{m,e-d} + x_{f,e-d} + y')$. By construction, the group G is capable if and only if H is capable. We now prove the theorem by constructing an isomorphism of rooted trees $\mathcal{D}_{d+k}(K_{e-k-d}) \cong \mathcal{D}_{d+k}(K_{e-k})$ which maps H onto G. Let

$$0 \leq j \leq d+k \quad \text{and} \quad i = 0, 1,$$

and denote by

$$\psi_j = \phi_{m,e-k-d,j} \colon \mathcal{Z}(m, e-k-d, j) \to \mathcal{Z}(m, e-k, j)$$

the isomorphism induced by multiplication by p. By Corollary 9.2, the j-step descendants of the group K_{e-k-id} correspond to $\Sigma_{m,e-k-id}(f)$-orbits on $\mathcal{Z}(m, e-k-id, j)$, and we now show that ψ_j maps a set of $\Sigma_{m,e-k-d}(f)$-orbit representatives onto a set of $\Sigma_{m,e-k}(f)$-orbit representatives. It follows from the assumptions and Lemma 9.8 that there is an automorphism $\alpha = \alpha(l, c, 1)$ of S which generates a Galois complement of $\Sigma_{m,s}(f)$ for all s in the given range, that is, $x_{f,s}^{\alpha} = x_{f,s}$ for all $e-k-d \leq s \leq e+d$.

If $x \in \mathcal{Z}(m, e-k-d, j)$, then Lemma 9.6 shows that

$$\psi_j(x^{\tilde{\alpha}}) = \psi_j(x^{\alpha} + x_{f,e-k-d+j}^{\alpha-1}) = \psi_j(x^{\alpha}) = \psi_j(x)^{\alpha} = \psi_j(x)^{\alpha} + x_{f,e-k+j}^{\alpha-1} = \psi_j(x)^{\tilde{\alpha}}$$

and, thus, the action of α on $\mathcal{Z}(m, e-k-d, j)$ and $\mathcal{Z}(m, e-d, j)$, respectively, is compatible with the isomorphism ψ_j.

Let N_i be the kernel of the projection from $\Sigma_{m,e-k-id}(f)$ to $\text{Aut}(P)$. By Lemma 9.8 and Theorem 5.14, we can choose n (and thus $m = n + d$ and $e = \mathfrak{e}_n$) large enough such that

$$N_i = \{\alpha(1, z, t) \mid z \in C_p(\theta) \times U_i, \, t \in T\} \quad \text{with} \quad U_i = \text{Stab}_{\mathcal{U}_p^{(2)}}(f + H_{e-k-id}),$$

and

$$U_1^{[p]} = U_0, \quad U_1 \leq \mathcal{U}_p^{(d)}, \quad \text{and} \quad U_0, U_1 \leq C_{\mathcal{U}_p^{(2)}}(H_v / H_{v+3d}) \text{ for all } v \geq 0.$$

Thus, N_i acts trivially on $\mathcal{Z}_l(m, e-k-id, j)$ and every $\alpha(1, z, 1) \in N_i$ acts trivially on $\mathcal{Z}_t(m, e-k-id, j)$. It follows from Lemma 9.7 that the linear actions of N_1 and N_0 are compatible with the isomorphism ψ_j.

Now we consider the affine actions of N_0 and N_1. Note that $x_{f,e-k-id+j}^{\beta-1} = 0$ for all $\beta = \alpha(1,\theta,t)$ with $t \in T$, cf. Section 9.2.1. If $\beta = \alpha(1,z,1)$ with $z \in U_1$, then $f^\beta - f = h$ lies in H_{e-k-d} and $f^{\beta^p} - f \equiv ph \mod H_{e-k+j}$ for all $0 \leq j \leq d+k$. This shows that

$$\psi_j(x^{\tilde{\beta}}) = \psi_j(x^\beta + x_{f,e-k-d+j}^{\beta-1}) = \psi_j(x + x_{h,e-k-d+j}) = \psi_j(x) + x_{ph,e-k+j}$$
$$= \psi_j(x)^{\beta^p} + x_{f,e-k+j}^{\beta^p-1} = \psi_j(x)^{\tilde{\beta}^p}$$

for all $x \in \mathcal{Z}(m, e-k-d, j)$. Thus, it follows from $U_1^{[p]} = U_0$ that ψ_j induces a bijection between the j-step descendants of K_{e-k-d} and the j-step descendants of K_{e-k}. By construction, these bijections give rise to a graph isomorphism from $\mathcal{D}_{d+k}(K_{e-k-d})$ to $\mathcal{D}_{d+k}(K_{e-k})$ which maps H onto G. □

9.11 Corollary. *The bound in Theorem 9.10 can be chosen as $n_0 = e_0 + 4p - 10$ if $p \geq 7$ and $n_0 = 32$ if $p = 5$, where e_0 is the integer given in Corollary 5.15.*

Proof. The proof of Theorem 9.10 requires that $\mathfrak{e}_n - 2d \geq \max\{-p^\star, e_0, p+1\}$, that is, $\mathfrak{e}_n \geq e_0 + 2d$ if $p > 5$, and $\mathfrak{e}_n \geq 28$ if $p = 5$. □

Computer experiments suggest that $n_0 \leq d^2/2 + 8d - 6$ if $p \geq 7$, cf. Remark 5.16. These bounds are not best possible, cf. Section 10.1 for $p = 5$.

We conclude this paragraph with an important corollary.

9.12 Corollary. *Let B be a capable group at depth $\mathfrak{e}_n$ in $\mathcal{T}_{n+d}$ and let A be the d-step parent of B. If $\mathrm{Aut}(A)$ is a p-group and $n \geq n_0$, then A is a periodic parent of B.*

Proof. We can assume that A and B lie on a path $K_0 \to K_1 \to \ldots$ in $\mathcal{T}_{n+d}$ defined by a surjective P-homomorphism $T \wedge T \to T$. If C is a descendant of A on this path, then Corollary 9.9 shows that $\mathrm{Aut}(C)$ is a p-group as well. Now Theorem 9.10 with $k = 0$ proves the assertion. □

9.3.3 The impact of the Galois complement

Let $p \equiv 5 \mod 6$ with $p > 5$, and $m = n + d$ with $n \geq n_0$ as in Corollary 9.11. Let $f \colon T \wedge T \to T$ be a surjective P-homomorphism, and denote by p_i the p'-part of $|\mathrm{Aut}(E_{m,i}(f))|$ for all $1 \leq i \leq \mathfrak{e}_m$. As already indicated in Theorem 9.10, the values of these p'-parts play an important role in the construction of a periodic parent of $E_{m,\mathfrak{e}_n}(f)$. We now describe a setback and support the conjecture based on computational investigations that there are infinitely many capable groups for which the d-step parent is not a periodic parent.

9.13 Remark. Let $e = \mathfrak{e}_n$, and we assume that $p_{e-d} = p_{e-d+1} > p_e = p_{e+1}$. Hence, we can choose generators α and β of Galois complements of $\Sigma_{m,e+1}(f)$ and $\Sigma_{m,e-d+1}(f)$, respectively, such that $\alpha = \beta^j$ for some $1 < j$ dividing the order of $\beta|_P$. Moreover, we assume that there is a twig tail vector $x \in \mathrm{T}(m, e+1)$ such that $x^\beta \neq x^{\alpha^i \alpha(1,1,t)}$ for all i and $t \in T$, cf. Lemma 9.5. With these assumptions, we prove that the d-step parent $E_{m,e-d}(f)$ of $E_{m,e}(f)$ is not a periodic parent of $E_{m,e}(f)$. For this purpose, we use Corollary 9.2 and show that the number of $\Sigma_{m,e-d}(f)$-orbits on $\mathcal{Z}(m, e-d, 1)$ is less than the number of $\Sigma_{m,e}(f)$-orbits on $\mathcal{Z}(m, e, 1)$. By the proof of Theorem 9.10, this holds if x and x^β do not lie in the same $\Sigma_{m,e}(f)$-orbit, that is, if $x^\beta \neq x^{\tilde\delta} = x^\delta + x_{f,e+1}^{\delta-1}$ for all $\delta \in \Sigma_{m,e}(f)$. Note that $x_{f,e+1}^{\delta-1}$ has a trivial twig component, see Lemma 7.16, and $\delta \in \Sigma_{m,e}(f)$ can be written as $\alpha^i \varepsilon$ for some $0 \leq i \leq d$ and $\varepsilon = \alpha(1,z,t)$ with $z \in C_p(\theta) \times \mathcal{U}_p^{(2)}$. By Lemma 9.5, the automorphism ε acts like $\alpha(1,1,t)$ on $\mathrm{T}(m,e+1)$. Hence, it follows from the choice of x that $x^\beta \neq x^{\tilde\delta}$ for all $\delta \in \Sigma_{m,e}(f)$.

We show in Theorem 10.1 that for $p=5$ the 4-step parent is always a periodic parent for large enough n. The proof relies on the fact that the $\mathbb{Z}_p[\theta]$-rank of $\mathrm{Hom}_P(T \wedge T, T)$ is 1 if $p=5$; that is, every surjective, liftable P-homomorphism $T \wedge T \to A_e$ is induced by some P-homomorphism $cF_2 : T \wedge T \to T$ with $c \in \mathfrak{p}^{-1} \setminus \mathfrak{p}^{-2}$, see Lemma 5.10.

In general, the $\mathbb{Z}_p[\theta]$-rank of $\mathrm{Hom}_P(T \wedge T, T)$ is $(p-3)/2$, see Corollary 5.6, and for $p>5$ there are surjective P-homomorphisms $f : T \wedge T \to T$ of the type

$$f = \sum_{\substack{2 \leq a \leq d/2 \\ a \neq a'}} c_a F_a + p^x F_{a'} \quad \text{with } x \geq 0.$$

Depending on the value of x, the summand $p^x F_{a'}$ has impact on the tail vector $x_{f,e}$, and hence on $\Sigma_{m,e}(f)$, only for sufficiently large e. Thus, for $p>5$, we see no argument why the assumptions of Remark 9.13 should not be satisfied for infinitely many homomorphisms f and values of m. In this case, Remark 9.13 points out that we cannot always choose the d-step parent as a periodic parent, and another construction of periodic parents is necessary. Theorem 9.10 is already a step ahead in solving this challenge.

10 5-groups of maximal class

As an application of the results obtained in this thesis, we now describe the structure of the coclass tree $\mathcal{T}$ of $\mathcal{G}(5)$ as conjectured in [41] and, recently, in [6]. In particular, we prove Conjecture 9.1 for $p = 5$ and show that the groups in the bodies of $\mathcal{T}$ can be described by finitely many parameterized presentations with at most two integer parameters. This is close to a positive answer of Problem 3 of Shalev's paper on finite p-groups [51], which asks for a classification of the 5-groups of maximal class. Throughout this chapter let $p = 5$.

10.1 The graph $\mathcal{G}(5)$

10.1 Theorem. *Let $n \geq 14$. If G is a capable group at depth $\mathfrak{e}_n = n - 4$ in $\mathcal{T}_{n+4}$, then the 4-step parent of G is a periodic parent.*

Proof. As mentioned in [41, p. 58], it follows from [3] and [26, 27] that the branch $\mathcal{B}_n$ has depth n for all $n \geq 4$. We say that $1 \leq e \leq n - 2$ is a ramification level of $\mathcal{B}_n$ if there is a group at depth e in $\mathcal{B}_n$ having more than one capable immediate descendant in $\mathcal{B}_n$. It follows from [28, Section 3] that the ramification levels of the body $\mathcal{T}_n$ are 1 and 5: There is one group at depth 1 in $\mathcal{T}_n$ having two capable immediate descendants and there are two groups at depth 5 in $\mathcal{T}_n$ having 5 and 2 capable immediate descendants, respectively. An analysis of the results in [28, Section 4] now shows that the p'-part of the automorphism group order of a capable group at depth greater than 5 in $\mathcal{T}_n$ is either 1 or 4, cf. [28, Lemma 4.3 & Theorem 4.4]. More precisely, if $G_1, \ldots, G_7$ are the seven capable groups at depth 6 in $\mathcal{T}_n$, then the p'-parts of the corresponding automorphism group orders are $1, \ldots, 1, 4$ and every capable descendant of G_i in $\mathcal{T}_n$ has an automorphism group order with the same p'-part as G_i for all $1 \leq i \leq 7$. Thus, if $n \geq 32$, then we can apply Theorem 9.10 to prove the periodicity of type 2. We continued the computations made in [6] and constructed the bodies $\mathcal{T}_2, \ldots, \mathcal{T}_{36}$ using the computer algebra system GAP and the GAP-package ANUPQ, see [44, 56]. These explicit constructions show that the periodic growth already starts at level $n = 14$. □

Figures 10.1–10.4 contain the bodies $\mathcal{T}_{14}, \ldots, \mathcal{T}_{17}$ of the coclass tree $\mathcal{T}$ of $\mathcal{G}(5)$. The notation is as explained in the Introduction, that is, a vertex labeled with an integer m stands for m terminal immediate descendants of the corresponding parent. It follows from Theorem 10.1 that these bodies and the periodicities of type 1 and 2 suffice to determine the structure of all bodies $\mathcal{T}_{14}, \mathcal{T}_{15}, \ldots$ completely.

It is conjectured in [28], [41], and [6, Conjecture IV] that the complete coclass graph $\mathcal{G}(5)$ can be described by two types of periodic patterns: First, the shaved branches $\mathcal{B}_n[n-1]$ and $\mathcal{B}_{n+4}[n-1]$ are conjectured to be isomorphic as rooted trees for all $n \geq 14$. Second, it is conjectured that the difference graph $\mathcal{B}_{n+4} \setminus \mathcal{B}_{n+4}[n-1]$, which consists of 5 levels of groups, is isomorphic to the difference graph $\mathcal{B}_n \setminus \mathcal{B}_n[n-5]$ for all $n \geq 14$.

Chapter 10. 5-groups of maximal class

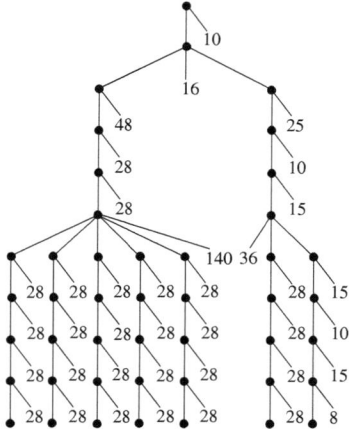

Figure 10.1: The body $\mathcal{T}_{14}$.

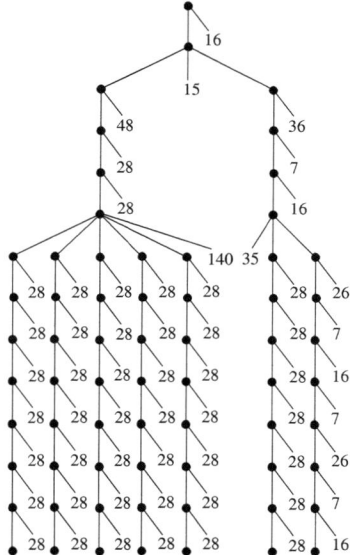

Figure 10.2: The body $\mathcal{T}_{17}$.

10.1. The graph $\mathcal{G}(5)$

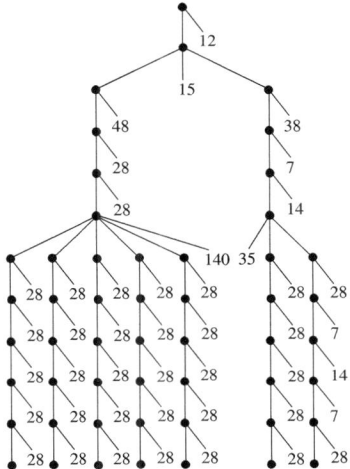

Figure 10.3: The body $\mathcal{T}_{15}$.

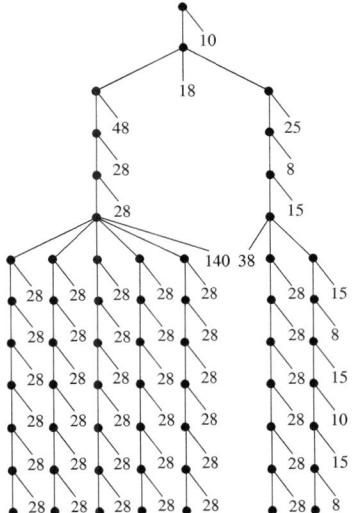

Figure 10.4: The body $\mathcal{T}_{16}$.

10.2 Periodicity classes

We now define **periodicity classes of type 2** for the groups in the bodies $\mathcal{T}_n$ with $36 \leq n \leq 39$ such that every group in $\mathcal{T}_m$ with $m \geq 36$ lies in exactly one periodicity class. Recall that $\mathcal{T}_n$ has depth $\mathfrak{e}_n = n - 4$.

If G has depth $e < \mathfrak{e}_{n-4}$ in $\mathcal{T}_n$ or if G has depth $e = \mathfrak{e}_{n-4}$ in $\mathcal{T}_n$ and G is terminal, then its periodicity class of type 2 is $\mathcal{P}_2(G) = \mathcal{P}(G)$, see Definition 8.5. By Theorem 8.6, the groups in $\mathcal{P}_2(G)$ can be described by a single parameterized presentation with one integer parameter.

If G has depth $e = \mathfrak{e}_{n-4} + k$ in $\mathcal{T}_n$ with $0 \leq k \leq 4$ and G is capable if $k = 0$, then we can assume that
$$G = E(\mathfrak{m}_{n,e} + x_{f,e} + v)$$
for some P-homomorphism $f \colon T \wedge T \to T$ and tail vector $v \in \mathcal{Z}(n, \mathfrak{e}_{n-4}, k)$. Recall that f is determined uniquely by its values on $\mathfrak{t}_1 \wedge \mathfrak{t}_3, \ldots, \mathfrak{t}_1 \wedge \mathfrak{t}_{d/2+1}$ and we can assume that $f(\mathfrak{t}_1 \wedge \mathfrak{t}_j) \in T_{\mathbb{Z}} \subseteq T$ for all $3 \leq k \leq d/2 + 1$. We define the periodicity class of G of type 2 as
$$\mathcal{P}_2(G) = \{ E(\mathfrak{m}_{n+4i, e+4j} + x_{f, e+4j} + \phi^j(v)) \mid 0 \leq i \text{ and } 0 \leq j \leq \lfloor \tfrac{\mathfrak{e}_{n+4i} - e}{4} \rfloor \}$$
where
$$\phi^j(v) \in \mathcal{Z}(n + 4i, \mathfrak{e}_{n-4} + 4j, k)$$
is the image of v under the "identity mapping" $\mathcal{Z}(n, \mathfrak{e}_{n-4}, k) \to \mathcal{Z}(n + 4i, \mathfrak{e}_{n-4}, k)$ and the isomorphism from $\mathcal{Z}(n + 4i, \mathfrak{e}_{n-4}, k)$ to $\mathcal{Z}(n + 4i, \mathfrak{e}_{n-4} + 4j, k)$ induced by multiplication by 5^j. By construction, $\mathcal{P}_2(G)$ is the union of infinitely many periodicity classes $\mathcal{P}(-)$ as defined in Definition 8.5 and, for all $i \geq 1$, the number of groups in $\mathcal{P}_2(G) \cap \mathcal{T}_{n+4i}$ is greater by one as the number of groups in $\mathcal{P}_2(G) \cap \mathcal{T}_{n+4(i-1)}$. By definition, every group in $\mathcal{P}_2(G)$ is defined by a consistent p.c.p. with generating set $\mathcal{S} \cup \mathcal{A} = \{\mathfrak{g}, \mathfrak{t}_1, \ldots, \mathfrak{t}_d, \mathfrak{a}_1, \ldots, \mathfrak{a}_d\}$. It follows from the results in Section 6.3.1, that each of the following three sets of relations
$$\{\mathcal{R}_{n+4i} \mid i \geq 0\}, \quad \{\mathcal{C}_{e+4j} \mid j \geq 0\}, \quad \text{and} \quad \{\mathcal{M}_{e+4j} \mid j \geq 0\}$$
can be described by a single parameterization with parameter i and j, respectively. By construction, the tail vectors in $\{\phi^j(v) \mid j \geq 0\}$ can be parameterized with parameter j, and Remark 6.11 implies that there exists $j_0 \in \mathbb{N}$ such that $\{x_{f, e+4j} \mid j \geq j_0\}$ can be described by a parameterization with parameter j. Note that $\mathfrak{m}_{n+4i, e+4j} = \mathfrak{m}_{n, e+4j}$ as tail vectors, see Section 7.1, and $\{\mathfrak{m}_{n, e+4j} \mid j \geq 0\}$ can be parameterized with parameter j.

In summary, this proves the following theorem.

10.2 Theorem. *The groups in the bodies of $\mathcal{T}$ can be described by finitely many parameterized presentations with at most two integer parameters.*

However, as indicated above, it is very technical to determine an explicit parameterized presentation describing the infinitely many groups in a periodicity class of type 2.

Moreover, there are 22218 groups in the bodies $\mathcal{T}_{36}, \ldots, \mathcal{T}_{39}$, that is, 22218 periodicity classes of type 2. In addition, the bodies $\mathcal{T}_2, \ldots, \mathcal{T}_{35}$ contain 71616 groups. This shows that an explicit classification by parameterized presentations still is very extensive.

10.2. Periodicity classes

The structure of the tree $\mathcal{T}$ has been investigated computationally in [41] and, recently, in [6]. Based only on these graph theoretic investigations, it is conjectured in [6] that all groups in the tree $\mathcal{T}$ can be partitioned into the set of 8399 groups lying in the branches $\mathcal{B}_2, \ldots, \mathcal{B}_{13}$, and into 8578 infinite *periodicity classes*, which partition the groups in $\mathcal{B}_{14}, \mathcal{B}_{15}, \ldots$ This and the results of Theorem 10.1 indicate that the number of periodicity classes of type 2 determined above seems not to be as small as possible.

A Appendix

A.1 Technical details

This section provides some straightforward, but technical, computations which are omitted in some proofs of the previous chapters.

A.1.1 Technical details of Remark 6.9

Recall that $n = \mathfrak{r}d + \mathfrak{i}$ with integers $\mathfrak{r} \geq 0$ and $1 \leq \mathfrak{i} \leq d$. We write $q = p^{\mathfrak{r}}$ and $k = d - \mathfrak{i} + 1$, and we consider the relation $t_i^{p^{\mathfrak{r}}} = w_{n,i}$ of the standard p.c.p. of S_n for $1 \leq i \leq k$. By construction, the element $w_{n,i}$ can be written as

$$w_{n,i} = t_{k+1}^{p^{\mathfrak{r}} e_{n,i,k+1}} \ldots t_d^{p^{\mathfrak{r}} e_{n,i,d}}$$

with $0 \leq e_{n,i,l} < p$ for all $1 \leq i \leq k$ and $k+1 \leq l \leq d$. It follows from $T_{n+d} = T_n^{[p]}$ that $e_{n+d,i,l} = e_{n,i,l}$ which shows that

$$w_{n+d,i} = t_{k+1}^{p^{\mathfrak{r}+1} e_{n,i,k+1}} \ldots t_d^{p^{\mathfrak{r}+1} e_{n,i,d}}$$

for all $1 \leq i \leq k$. The word $t_1^{l_{n,1}} \ldots t_d^{l_{n,d}}$ is the normalized word of $t_1^{-1} \ldots t_d^{-1}$ in S_n and therefore

$$l_{n,i} = p^{\mathfrak{r}} - 1 \qquad (1 \leq i \leq k) \quad \text{and}$$
$$l_{n,j} = f_{n,j} \bmod p^{\mathfrak{r}+1} \qquad (k+1 \leq j \leq d)$$

with

$$f_{n,j} = p^{\mathfrak{r}}(p + \sum_{i=1}^{k}(p - e_{n,i,j})) - 1.$$

This already shows that $l_{n,i}$ is congruent to -1 modulo $p^{\mathfrak{r}}$ for all i. If $1 \leq i \leq k$, then

$$p(l_{n,i} + 1) - 1 = l_{n+d,i}.$$

If $k+1 \leq j \leq d$, then $p(l_{n,j} + 1) - 1 \equiv f_{n+d,j} \equiv l_{n+d,j} \bmod p^{\mathfrak{r}+2}$ and

$$l_{n+d,j} = p(l_{n,j} + 1) - 1$$

follows from $0 \leq p(l_{n,j} + 1) - 1 < p^{\mathfrak{r}+2}$. It is easy to proceed with an inductive argument.

A.1.2 The unit $1 + \theta + \ldots + \theta^i$ with $0 \leq i \leq d-1$

In this paragraph, we investigate the ω- and θ-part of the unit

$$u_i = 1 + \theta + \ldots + \theta^i$$

for $0 \leq i \leq d-1$. Recall that $\omega \in \mathbb{Z}_p$ is a primitive $(p-1)$-th root of unity and $\omega \bmod p$ is a generator of $\mathbb{F}_p^\star$.

Lemma. *The unit* $u_i = 1 + \theta + \ldots + \theta^i$ *with* $0 \leq i \leq d-1$ *can be written as*
$$u_i = (i+1)\theta^{i(d/2+1)}(1+s_i) \quad \text{with } s_i \in \mathfrak{p}^2$$
and $u_i = \omega^{j_i}\theta^{i(d/2+1)}(1+\tilde{s}_i)$ *with* $\tilde{s}_i \in \mathfrak{p}^2$ *and* $0 \leq j_i \leq d-1$ *such that* $\omega^{j_i} \equiv i+1 \bmod p$.

Proof. By definition, $u_i = (\theta^{i+1} - 1)/(\theta - 1)$, and θ and θ^{i+1} are primitive roots of unity. If $\psi = \theta^{i+1}$ and $j = (i+1)^{-1} \bmod p$, then $(\psi^j - 1)/(\psi - 1) = (\theta - 1)/(\theta^{i+1} - 1) \in \mathbb{Z}_p[\psi]$ and it follows from $\mathbb{Z}_p[\psi] = \mathbb{Z}_p[\theta]$ that u_i is a unit. The element u_0 can be written as $u_0 = 1\theta^0(1+s_0)$ with $s_0 = 0 \in \mathfrak{p}^2$, and we proceed by induction. Thus, we assume that $u_i = (i+1)\theta^{i(d/2+1)}(1+s_i)$ for some $s_i \in \mathfrak{p}^2$ and $i < d-2$, and we have to show that there exists $s_{i+1} \in \mathfrak{p}^2$ with
$$u_{i+1} = 1 + \theta u_i = (i+2)\theta^{(i+1)(d/2+1)}(1+s_{i+1}),$$
which is equivalent to $1 + (i+1)\theta^{i(d/2+1)+1} - (i+2)\theta^{(i+1)(d/2+1)} \in \mathfrak{p}^2$. Another induction shows that $\theta^h \equiv h\theta - h + 1 \bmod (\mathfrak{p}^2, +)$ for all $h \geq 0$; that is,
$$1 + (i+1)\theta^{i(d/2+1)+1} - (i+2)\theta^{(i+1)(d/2+1)} \equiv -(i+1)p\theta + (i+1)p \equiv 0 \bmod (\mathfrak{p}^2, +),$$
which proves the first assertion of the lemma. For $0 \leq i \leq d-1$ there is $0 \leq j_i \leq d-1$ with $\omega^{j_i} \equiv i+1 \bmod p$; that is, $\omega^{j_i} = i+1+pz$ for some $z \in \mathbb{Z}_p$. This shows that $u_i = (i+1)\theta^{i(d/2+1)}(1+s_i)$ can be written as
$$u_i = \omega^{j_i}\theta^{i(d/2+1)}(1+\tilde{s}_i)$$
with $\tilde{s}_i = \omega^{-j_i}(i+1)(1+s_i) - 1 \in \mathfrak{p}^2$. □

A.1.3 Proof of Lemma 6.17

We now provide a proof for Lemma 6.17.

Lemma. *If A is an $S_\mathbb{Z}$-module with $T_\mathbb{Z}$ acting trivially on A, then*
$$H^2(S_\mathbb{Z}, A) \cong (A^P)^3 \times A^{d/2-1}.$$

Proof. Let $\langle \mathcal{D} \mid \mathcal{B} \rangle$ be a consistent p.c.p. of A and let $\mathcal{M}$ be the set of conjugate relations in $\mathcal{S} \cup \mathcal{D}$ describing the $S_\mathbb{Z}$-module structure on A. Let $x = (x_{i,j})$ be a list of elements in A defining a polycyclic presentation $\mathcal{E}'(x)$ with generators $\mathcal{S} \cup \mathcal{D}$ and relations
$$\{\mathfrak{g}^p = x_{0,0}, \quad \mathfrak{t}_k^{\mathfrak{t}_l} = \mathfrak{t}_k x_{l,k}, \quad \mathfrak{t}_j^{\mathfrak{g}} = \mathfrak{t}_{j+1} x_{0,j}, \quad \mathfrak{t}_d^{\mathfrak{g}} = \mathfrak{t}_1^{-1} \ldots \mathfrak{t}_d^{-1} x_{0,d} \mid l < k \text{ and } j < d\} \cup \mathcal{B} \cup \mathcal{M}.$$

We now use consistency checks as in Theorem 4.2 to impose necessary and sufficient conditions on x to define a consistent presentation $\mathcal{E}'(x)$.

First, $x_{0,0}^{\mathfrak{g}} = x_{0,0}$ follows from $\mathfrak{g}(\mathfrak{g}^p) = \mathfrak{g}x_{0,0}$ and $(\mathfrak{g}^p)\mathfrak{g} = \mathfrak{g}x_{0,0}^{\mathfrak{g}}$. If $0 < j < i < d$, then $x_{j+1,i+1} = x_{j,i}^{\mathfrak{g}}$ follows from
$$\mathfrak{t}_i(\mathfrak{t}_j\mathfrak{g}) = \mathfrak{g}\mathfrak{t}_{j+1}\mathfrak{t}_{i+1}x_{0,i}x_{0,j}x_{j+1,i+1} \quad \text{and} \quad (\mathfrak{t}_i\mathfrak{t}_j)\mathfrak{g} = \mathfrak{g}\mathfrak{t}_{j+1}\mathfrak{t}_{i+1}x_{j,i}^{\mathfrak{g}}x_{0,i}x_{0,j}.$$

If $0 < i < d$, then $x_{i,d}^{\mathfrak{g}} \prod_{k=1}^{i} x_{k,i+1}^{-1} = \prod_{k=i+2}^{d} x_{i+1,k}^{-1}$ is a consequence of
$$(\mathfrak{t}_d\mathfrak{t}_i)\mathfrak{g} = \mathfrak{g}\mathfrak{t}_{i+1}\mathfrak{t}_1^{-1}\ldots\mathfrak{t}_d^{-1}x_{0,d}x_{0,i}x_{i,d}^{\mathfrak{g}} = \mathfrak{g}\mathfrak{t}_1^{-1}\ldots\mathfrak{t}_i^{-1}\mathfrak{t}_{i+2}^{-1}\ldots\mathfrak{t}_d^{-1}x_{0,i}x_{0,d}x_{i,d}^{\mathfrak{g}} \prod_{k=1}^{i} x_{k,i+1}^{-1}$$
and $\mathfrak{t}_d(\mathfrak{t}_i\mathfrak{g}) = \mathfrak{g}\mathfrak{t}_1^{-1}\ldots\mathfrak{t}_i^{-1}\mathfrak{t}_{i+2}^{-1}\ldots\mathfrak{t}_d^{-1}x_{0,i}x_{0,d}\prod_{k=i+2}^{d}x_{i+1,k}^{-1}$.

A.1. Technical details

For all $k < j < i$ the equation $\mathfrak{t}_i(\mathfrak{t}_j\mathfrak{t}_k) = \mathfrak{t}_k\mathfrak{t}_j\mathfrak{t}_i x_{k,j} x_{k,i} x_{j,i} = (\mathfrak{t}_i\mathfrak{t}_j)\mathfrak{t}_k$ yields no further conditions. Finally, it follows from $\mathfrak{t}_i(\mathfrak{g}^p) = \mathfrak{t}_i x_{0,0}$ and

$$\begin{aligned}
(\mathfrak{t}_i\mathfrak{g})\mathfrak{g}^{p-1} &= \mathfrak{g}\mathfrak{t}_{i+1}\mathfrak{g}^{p-1} x_{0,i}^{\mathfrak{g}^{p-1}} = \ldots = \mathfrak{g}^{d-i+1}\mathfrak{t}_1^{-1}\cdots\mathfrak{t}_d^{-1}\mathfrak{g}^i \underbrace{\prod_{k=i}^d x_{0,k}^{\mathfrak{g}^{p+i-k-1}}}_{=s_1} \\
&= \mathfrak{g}^{d-i+2}\mathfrak{t}_2^{-1}\cdots\mathfrak{t}_d^{-1}\mathfrak{t}_d\cdots\mathfrak{t}_1\mathfrak{g}^{i-1}s_1 \underbrace{(\prod_{k=1}^d x_{0,k}^{-1})^{\mathfrak{g}^{i-1}}}_{=s_2} \\
&= \mathfrak{g}^{d-i+2}\mathfrak{t}_1\mathfrak{g}^{i-1}s_1s_2 = \ldots = \mathfrak{g}^p\mathfrak{t}_i s_1 s_2 \underbrace{\prod_{k=1}^{i-1} x_{0,i}^{\mathfrak{g}^{i-k-1}}}_{=s_3} = \mathfrak{t}_i x_{0,0} s_1 s_2 s_3
\end{aligned}$$

for all $1 \leq i \leq d$ that $\prod_{k=1}^d x_{0,k}^{\mathfrak{g}^{p-k}-1} = 1$. Thus, $\mathcal{E}'(x)$ is consistent if and only if x satisfies

(1) $x_{i,j}^{\mathfrak{g}} = x_{i+1,j+1}$ for $0 < i < j < d$,

(2) $x_{i,d}^{\mathfrak{g}} \prod_{k=1}^i x_{k,i+1}^{-1} = \prod_{k=i+2}^d x_{i+1,k}^{-1}$ for $0 < i < d$,

(3) $x_{0,0} \in A^P$, and

(4) $\prod_{k=1}^d x_{0,k}^{\mathfrak{g}^{p-k}-1} = 1$.

Equation (1) shows that $x_{i,j} = x_{1,j-i+1}^{\mathfrak{g}^{i-1}}$ for all $1 < i < j \leq d$ and Equation (4) leads to

$$x_{0,d} \equiv \prod_{k=1}^{d-1} (x_{0,k}^{-1})^{\beta_k} \mod A^P \text{ with } \beta_k = 1 + \mathfrak{g} + \ldots + \mathfrak{g}^{d-k}.$$

Using Equation (1), Equation (2) can be translated to $\prod_{k=2}^{d-i+1} x_{1,k} = \prod_{k=2}^{i+1} x_{1,k}^{\mathfrak{g}^{1-k}}$ for $0 < i < d$ which is equivalent to

$$x_{1,i} = x_{1,d-i+3}^{-\mathfrak{g}^{i-1}} \text{ for } 3 \leq i \leq d \text{ and } x_{1,2}^{\mathfrak{g}} = \prod_{k=2}^d x_{1,k}^{\mathfrak{g}^{1-k}}.$$

Thus $x_{1,3}, \ldots, x_{1,d/2+1}$ can be chosen arbitrarily and determine

$$x_{1,i} = x_{1,d-i+3}^{-\mathfrak{g}^{i-1}} \text{ for } d/2+2 \leq i \leq d \text{ and } x_{1,2} \equiv \prod_{k=3}^{d/2+1} x_{1,k}^{\alpha_k} \mod A^P$$

with $\alpha_k = \mathfrak{g} + \ldots + \mathfrak{g}^{p-k}$. This finally shows that $\mathcal{Z}(S_{\mathbb{Z}}, A) \cong (A^P)^3 \times A^{d-1} \times A^{d/2-1}$.

We now determine $\mathcal{B}(S_{\mathbb{Z}}, A)$ and consider a generating set $\{\mathfrak{g}y_0, \mathfrak{t}_1 y_1, \ldots, \mathfrak{t}_d y_d\}$ with $y_0, \ldots, y_d \in A$ of a complement to A in the group defined by $\mathcal{E}'(x)$. The equations

$$\mathfrak{t}_j y_j = \mathfrak{t}_j y_j^{\mathfrak{t}_i y_i} = \mathfrak{t}_j x_{i,j} y_j \text{ and } 1 = (\mathfrak{g} y_0)^p = x_{0,0} y_0^{1+\mathfrak{g}+\ldots+\mathfrak{g}^d}$$

imply that $x_{i,j} = 1$ for $1 \leq i < j \leq d$ and $x_{0,0} = 1$. Moreover, $x_{0,i} = y_{i+1} y_i^{-\mathfrak{g}}$ for $1 \leq i < d$ and $x_{0,d} = y_1^{-1} \cdots y_{d-1}^{-1} y_d^{1-\mathfrak{g}}$. This shows that $x_{0,1}, \ldots, x_{0,d-1}$ can be chosen arbitrarily and the value of $x_{0,d}$ is determined by

$$x_{0,d} = y_1^{-1} \cdots y_{d-1}^{-1} y_d^{1-\mathfrak{g}} = \prod_{k=1}^{d-1} (x_{0,k}^{-1})^{\beta_k}.$$

Therefore, $\mathcal{B}(S_{\mathbb{Z}}, A) \cong A^{d-1}$ and the lemma is proved. □

A.1.4 Technical details of Lemmas 6.25 and 7.1

Recall that $n = \mathfrak{r}d + \mathfrak{i}$ with integers $\mathfrak{r} \geq 0$ and $1 \leq \mathfrak{i} \leq d$. We write $q = p^{\mathfrak{r}}$ and $k = d - \mathfrak{i} + 1$. If $x = x_{f \circ \gamma}$ is defined by $f \in \mathrm{Hom}_P(T_{\mathbb{Z},n}, A_e)$, then, by the proof of Lemma 6.25, the possibly non-trivial tails of $x = (x_{i,j})$ are

$$x_{i,i} = f(w_i) \quad \text{with} \quad w_i = \mathfrak{t}_i^q w_{n,i}^{-1} \in T_{\mathbb{Z},n} \quad \text{if } 1 \leq i \leq k \text{ and}$$
$$x_{j,j} = f(w_j) \quad \text{with} \quad w_j = \mathfrak{t}_j^{pq} \in T_{\mathbb{Z},n} \quad \text{if } k+1 \leq j \leq d$$

where the inverses of $w_{n,1}, \ldots, w_{n,d-\mathfrak{i}+1}$ are computed in $T_{\mathbb{Z}}$. The tail $x_{0,d}$ is

$$x_{0,d} = f(w) \quad \text{with} \quad w = \mathfrak{t}_1^{-l_{n,1}-1} \ldots \mathfrak{t}_d^{-l_{n,d}-1} \in T_{\mathbb{Z},n}.$$

The group $E(x)$ satisfies $\mathfrak{g}^p = 1$ and, hence,

$$(\mathfrak{g}\mathfrak{t}_1)^p = \mathfrak{t}_1^{l_{n,1}+1} \ldots \mathfrak{t}_d^{l_{n,d}+1} x_{0,d}.$$

We now show that $\mathfrak{t}_1^{l_{n,1}+1} \ldots \mathfrak{t}_d^{l_{n,d}+1} = x_{0,d}^{-1}$ in $E(x)$ which proves Lemma 7.1. It follows as in Section A.1.1 that

$$l_{n,i} = q - 1 \quad (1 \leq i \leq k) \quad \text{and}$$
$$l_{n,j} = f_{n,j} \bmod pq \quad (k+1 \leq j \leq d)$$

with

$$f_{n,j} = q(p + \sum_{i=1}^{k}(p - e_{n,i,j})) - 1$$

where $0 \leq e_{n,i,l} < p$ such that $w_{n,i} = \mathfrak{t}_{k+1}^{qe_{n,i,k+1}} \ldots \mathfrak{t}_d^{qe_{n,i,d}}$ for all $1 \leq i \leq k$. If we write

$$l_{n,j} = f_{n,j} - pqd_{n,j} \quad (k+1 \leq j \leq d)$$

for certain $d_{n,j} \in \mathbb{Z}$, then $E(x)$ satisfies

$$\mathfrak{t}_1^{l_{n,1}+1} \ldots \mathfrak{t}_d^{l_{n,d}+1} = x_{1,1} \ldots x_{k,k} \prod_{j=k+1}^{d} x_{j,j}^{1-d_{n,j}+k}$$

and

$$\begin{aligned}
x_{0,d} &= f(\mathfrak{t}_1^{-l_{n,1}-1} \ldots \mathfrak{t}_d^{-l_{n,d}-1}) \\
&= f(\mathfrak{t}_1^{-q} w_{n,1} \ldots \mathfrak{t}_k^{-q} w_{n,k} \prod_{j=k+1}^{d} \mathfrak{t}_j^{(d_{n,j}-(k+1))pq}) \\
&= x_{1,1}^{-1} \ldots x_{k,k}^{-1} \prod_{j=k+1}^{d} x_{j,j}^{-1+d_{n,j}-k}.
\end{aligned}$$

This proves Lemma 7.1 and completes the proof of Lemma 6.25.

A.2 Coclass conjectures

The coclass conjectures, Conjectures A – E, together with the theory developed for their proofs can be seen as the fundamentals of coclass theory and, therefore, we give a brief survey of these conjectures in this section. We state them as theorems since they all have been proved now. It is the case (though not obviously) that Theorem A implies all the others. Let p be a prime and let r be a positive integer in this section.

Theorem A. *There is a function $f(p,r)$ such that every p-group of coclass r has a normal subgroup of nilpotency class at most 2 and index at most $f(p,r)$.*

Theorem B. *There is a function $g(p,r)$ such that every p-group of coclass r has derived length at most $g(p,r)$.*

Theorem C. *Every pro-p group of coclass r is solvable.*

Theorem D. *There are only finitely many isomorphism types of infinite pro-p groups of coclass r.*

Theorem E. *There are only finitely many isomorphism types of infinite solvable pro-p groups of coclass r.*

By [7, p. 267], there is a one-to-one correspondence between the isomorphism types of infinite pro-p groups of coclass r and the coclass trees in $\mathcal{G}(p,r)$, where a coclass tree is associated with an inverse limit of its mainline. Hence, Theorem D is equivalent to

Theorem D'. *The coclass graph $\mathcal{G}(p,r)$ contains only finitely many coclass trees.*

It follows from [29, Corollary 11.2.3] that there are only finitely many groups in $\mathcal{G}(p,r)$ which are not contained in any coclass tree.

A.2.1 On the proof of Theorem A

A large number of papers contributed to the ultimate proof of Theorem A, and we only mention a few highlights in the following historical abstract, which is taken from [7, pp. 265]. It is in chronological order and we refer to [7] and [29] for background and further references.

(a) Leedham-Green & Newman [30] defined coclass, and proposed a classification of p-groups of fixed coclass. Their program was expressed in a series of five conjectures.

(b) Leedham-Green [31] proved that every p-group of finite coclass is p-adic analytic.

(c) Donkin [8] proved that if $p \geq 5$, then every p-adic analytic pro-p group of finite coclass is solvable. Together with (b) this establishes Theorem C for $p \geq 5$.

(d) Leedham-Green [32] completes the proof of Theorem A for $p \geq 5$.

(e) Shalev and Zel'manov [49] give an elementary self-contained proof of Donkin's theorem (c), valid for all primes p.

(f) Shalev [50] gives a 'constructive' self-contained proof of Theorem A, valid for all primes.

Hence, the proof of Theorem A has a long history and there are two independent proofs by Leedham-Green [32] and Shalev [50]. A *featured online review* of these two papers was written by Mann [33]. We now quote an excerpt of this review which emphasizes on the different ingredients of the two proofs. Introductorily, Mann wrote:

> These papers represent a significant advance in the theory of finite p-groups. Since the number of p-groups is vast, and includes many strange ones, there is a widely shared belief that their structure is chaotic. Some years ago C. R. Leedham-Green and M. F. Newman suggested a way of bringing order into this chaos (an extensive bibliography and historical background are available in the papers themselves). If G has order p^n and nilpotency class c, then Leedham-Green and Newman called $n - c$ the coclass of G, and considered the coclass as the main invariant of G. Based on some evidence (mostly from the theory of groups of maximal class, i.e. coclass 1), and on profound intuition, they made several bold conjectures, the strongest of which states that a p-group of coclass r contains a subgroup of class 2 (abelian if $p = 2$) whose index is bounded in terms of r (and p). Thus p-groups of small coclass are virtually of class 2. According to a reported saying of N. Blackburn, this conjecture, once proved, will be "the first general theorem of the theory of p-groups". Both of the present papers contain a proof. The proof by Leedham-Green, though published later, was the first one, and was available to Shalev. It is the culmination of work by Leedham-Green, S. McKay, W. Plesken, and S. Donkin. (A special case was done by the reviewer, based on work of McKay.) Because of its reliance on Donkin's work, which applies the classification of simple p-adic Lie algebras, this proof originally held only for $p > 3$, but this restriction can be removed now, by substituting Donkin's paper with one of A. Shalev and E. I. Zel'manov.

Concerning the proof of Leedham-Green, Mann continues:

> The main ingredients in Leedham-Green's proof are the concepts of uniserial action and of settled groups, a "reduction" from finite p-groups to (infinite) pro-p-groups, and the application of Lie methods. In a group G (all groups from now on are p-groups) of coclass r, at most r factors of the lower central series have order greater than p, and that means that G acts uniserially on a large chunk of its lower central series. The author calls a group G settled if it acts uniserially on a normal subgroup N that is both large enough and has, in some sense, a suitable power structure. By playing this power structure against the commutator structure, the author defines a certain Lie ring for each settled group.
>
> The proof proper starts by assuming that the result does not hold. Then there exists an infinite set of groups of the same coclass r, which together exhibit the failure of the theorem. An inverse limit argument constructs a pro-p-group P of coclass r. By a previous paper of Leedham-Green, P is p-adic analytic. This enables us to apply the result of Donkin (or of Shalev and Zel'manov), together with Leedham-Green's previous results. This yields a detailed structure of P, and in particular shows that it is settled. It follows that only finitely many groups of coclass r can fail to be settled. Thus the groups in the infinite set above can be taken as settled. Now another inverse limit construction is applied to build from the Lie rings associated to these groups an infinite Lie ring on which P acts uniserially. This contradicts Donkin's result and ends the proof.
>
> The proof for the case $p = 2$, in which the result is stronger, is actually shorter. The whole proof, including previous stages, is surprisingly relatively short and simple, especially as Donkin's work, the only "non-elementary" ingredient, can now be eliminated. Moreover, Leedham-Green, not content with his achievements so far, next discusses the exact structure of groups of a given coclass. He introduces a notion of constructible groups. The definition is somewhat complicated, and involves cohomology. We will just say that constructible groups are obtained by twisting finite quotients of p-adic space

groups. The paper ends with the following remarkable theorem: *A group G of coclass r contains a normal subgroup N of bounded order (in terms of r and p) such that G/N is constructible.*

The approach of Shalev is described by Mann as follows:

> Now to Shalev's paper: its main advantages are, first, that it gives a largely self-contained elementary proof, and second, that it gives explicit bounds, which, while they are large, are claimed by the author to be "quite realistic". This is possible because he avoids inverse limit constructions. Rather, he applies the theory of powerful p-groups, as developed by A. Lubotzky and the reviewer. This theory can be considered as the finite analogue of Lazard's theory of p-adic analytic groups. Again using uniserial actions, it is first established that a group of coclass r contains a powerful subgroup of bounded index. This is analogous, and even equivalent, to Leedham-Green's result that a pro-p-group of finite coclass is p-adic analytic. Since powerful groups have nice power and commutator structure, it is again possible to construct from an assumed minimal counterexample an appropriate Lie ring. This ring is modified several times, ending with a perfect Lie algebra L of characteristic p with a derivation D such that $D^{p-1} = 0$. But a result of N. Jacobson on Engel Lie algebras shows that L is nilpotent, a contradiction. Shalev ends by giving the proof for $p = 2$, which here is longer than for odd p, by proving that most groups are settled, a result that here is a corollary, rather than preliminary, to the main result, and deriving some other results about the structure of groups of a given coclass.

A.2.2 On the proofs of Theorems B – E

We now sketch the proofs of Theorems B – E and provide further references. The proofs of Theorem B and C are from [29, Corollary 6.4.6 & Theorem 7.4.2]. The proof of Theorem D can be found in [7, Theorem 10.2].

Proof of Theorem B. Let G be a p-group of coclass r and denote by $G^{(1)} = G$ and $G^{(i)} = [G^{(i-1)}, G^{(i-1)}]$ for $i \geq 2$ the terms of the derived series of G. Since $G^{(i)} \leq \gamma_{2^i}(G)$, it follows from the proof of Theorem A, see [29, Section 6.4], that G has derived length at most $i + 2$ if $2^i \geq 2(p-1)p^{r-1} - 2$ for p odd, and if $2^{i+1} \geq 3.2^{r+1} - 1$ for $p = 2$. □

Proof of Theorem C. Let G be an infinite pro-p group of coclass r, that is, there exists an integer i_0 such that $G/\gamma_i(G)$ is a finite p-group of coclass r for all $i \geq i_0$. By Theorem B, the group $G/\gamma_i(G)$ has derived length at most $g(p, r)$, where $g(p, r)$ is independent of i. Hence $G^{(g(p,r))} \leq \bigcap \gamma_i(G) = \{1\}$, and so G is solvable. □

Proof of Theorem D. By [7, Chapter 10], an infinite pro-p group G of coclass r has an open normal subgroup $A \cong \mathbb{Z}_p^k$, where $k = (p-1)p^s$ for some $s < r$ if p is odd, $k = 2^s$ for some $s < r + 1$ if $p = 2$. Moreover, G/A has coclass r and $[G : A] = p^{r+p^r}$ if p is odd, $[G : A] = 2^{r+(r+1)2^{r+1}}$ if $p = 2$, see [7, Theorem 10.1]. As indicated in [7, Section 10.4], the proof of this assertion relies on Theorem C. By [7, Theorem 5.8], there are only finitely many isomorphism types of extensions of the pro-p group $\mathbb{Z}_p^k$ by a finite p-group. □

Proof of Theorem E. Obviously, this is a consequence of Theorem D. □

Bibliography

[1] H. U. Besche and B. Eick, *Construction of finite groups*, J. Symb. Comput. **27** (1999), 387 – 404.

[2] H. U. Besche, B. Eick, and E. A. O'Brien, *A millennium project: constructing small groups*, Internat. J. Algebra Comput., **12** (2002), 623 – 644.

[3] N. Blackburn, *On a special class of p-groups*, Acta Math. **100** (1958), 45 – 92.

[4] K. Dekimpe and B. Eick, *Computational aspects of group extensions and their applications in topology*, Exp. Math. **11** No. 2 (2002), 183 – 200.

[5] H. Dietrich, *Periodic patterns in the graph of p-groups of maximal class*, submitted to Journal of Group Theory (2008).

[6] H. Dietrich, B. Eick, and D. Feichtenschlager, *Investigating p-groups by coclass with GAP*, Contemp. Mathematics, Amer. Math. Soc. **470** (2008), 45 – 61.

[7] J. D. Dixon, M. P. F. du Sautoy, A. Mann, and D. Segal, Analytic pro-p-groups. 2nd edition, Cambridge University Press, 2003.

[8] S. Donkin, *Space groups and groups of prime-power order, VIII. Pro-p groups of finite coclass and p-adic Lie algebras*, J. Algebra **111** (1987), 316 – 342.

[9] M. P. F. du Sautoy, *Counting p-groups and nilpotent groups*, Inst. Hautes Etudes Sci. Publ. Math. **92** (2001), 63 – 112.

[10] B. Eick, *Algorithms for polycyclic groups*, Habilitationsschrift, Universität Kassel, 2001.

[11] B. Eick, *Determination of the uniserial space groups with a given coclass*, J. London Math. Soc. **71** (2005), 622 – 642.

[12] B. Eick, *Automorphism groups of 2-groups*, J. Algebra **300** (2006), 91 – 101.

[13] B. Eick, *Schur multiplicators of finite p-groups with fixed coclass*, Israel J. Math. **166** (2008), 157 – 166.

[14] B. Eick and C. R. Leedham-Green, *On the classification of prime-power groups by coclass*, Bull. London Math. Soc. **40** (2008), 274 – 288.

[15] B. Eick, C. R. Leedham-Green, M. F. Newman, and E. A. O'Brien *Classification of groups with prime-power order by coclass II*, preprint, (2009).

[16] B. Eick, A. C. Niemeyer, and O. Panaia, *A polycyclic quotient algorithm*, preprint, (2008).

[17] G. A. Fernández-Alcober, *The exact lower bound for the degree of commutativity of a p-group of maximal class*, J. Algebra, **174** (1995), 523 – 530.

[18] F. Q. Gouvêa, *p*-adic numbers – an introduction. 2nd edition, Springer, 2003.

[19] H. Hasse, Number theory. Reprint of the 1980 ed.. Springer, 2002.

[20] G. Higman, *Enumerating p-groups I: Inequalities*, Proc. London Math. Soc. (3) **10** (1960), 24 – 30.

[21] G. Higman, *Enumerating p-groups II: Problems whose solution is PORC*, Proc. London Math. Soc. (3) **10** (1960), 566 – 582.

[22] D. F. Holt, B. Eick, and E. A. O'Brien, Handbook of computational group theory. Discrete Math. Appl., CRC Press, 2005.

[23] B. Huppert, Endliche Gruppen I. Springer Heidelberg, 1967.

[24] A. Jaikin-Zapirain and A. Vera-López, *On the use of the Lazard correspondence in the classification of p-groups of maximal class*, J. Algebra **228** no. 2 (2000), 477 – 490.

[25] C. R. Leedham-Green and S. McKay, *On p-groups of maximal class I*, Quart. J. Math. Oxford (2) **27** (1976), 297–311.

[26] C. R. Leedham-Green and S. McKay, *On p-groups of maximal class II*, Quart. J. Math. Oxford (2) **29** (1978), 175 – 186.

[27] C. R. Leedham-Green and S. McKay, *On p-groups of maximal class III*, Quart. J. Math. Oxford (2) **29** (1978), 281 – 299.

[28] C. R. Leedham-Green and S. McKay, *On the classification of p-groups of maximal class*, Quart. J. of Math. Oxford **35** (1984), 293 – 304.

[29] C. R. Leedham-Green and S. McKay, The structure of groups of prime power order. London Mathematical Society Monographs, Oxford Science Publications, 2002.

[30] C. R. Leedham-Green and M. F. Newman, *Space groups and groups of prime-power order I*, Archiv der Mathematik **35** (1980), 193 – 203.

[31] C. R. Leedham-Green, *Pro-p groups of finite coclass*, J. London Math. Soc. **50** (1994), 43 – 48.

[32] C. R. Leedham-Green, *The structure of finite p-groups*, J. London Math. Soc. **50** (1994), 49 – 67.

[33] A. Mann, a featured online review for the papers [32] and [50] (1994), available at www.ams.org/mathscinet/search/publdoc.html?pg1=MR&r=1&s1=1258908&vfpref=html

[34] U. Martin, *Almost all p-groups have automorphism group a p-group*, Bull. Amer. Math. Soc. **15** no. 1 (1986), 78 – 82.

[35] R. J. Miech, *Metabelian p-groups of maximal class*, Trans. Amer. Math. Soc. **152** (1970), 331 – 373.

[36] R. J. Miech, *Some p-groups of maximal class*, Trans. Amer. Math. Soc. **189** (1974), 1 – 47.

[37] R. J. Miech, *Counting commutators*, Trans. Amer. Math. Soc. **189** (1974), 49 – 61.

[38] R. J. Miech, *The metabelian p-groups of maximal class II*, Trans. Amer. Math. Soc. **236** (1978), 93 – 119.

[39] R. J. Miech, *The metabelian p-groups of maximal class II*, Trans. Amer. Math. Soc. **272** no. 2 (1982), 465 – 474.

[40] J. Neukirch, Algebraische Zahlentheorie. Springer, 1992.

[41] M. F. Newman, *Groups of prime-power order*, Groups-Canberra 1989, Lecture notes in mathematics, vol. **1456**, Springer (1990), 49 – 62.

[42] M. F. Newman and E. A. O'Brien, *Classifying 2-groups by coclass*, Trans. Amer. Math. Soc. **351** (1999), 131 – 169.

[43] M. F. Newman, E. A. O'Brien, and M. R. Vaughan-Lee, *Groups and nilpotent Lie rings whose order is the sixth power of a prime*, J. Algebra **278** (2004), 383 – 401.

[44] E. A. O'Brien, ANUPQ- *the anu p-quotient algorithm*, 1990, available in MAGMA and as GAP package.

[45] E. A. O'Brien and M. R. Vaughan-Lee, *The groups of order p^7 for odd prime p*, J. Algebra **292** (2005), 243 – 258.

[46] A. M. Robert, A course in p-adic analysis. Springer, 2000.

[47] D. J. S. Robinson, A course in the theory of groups. Springer-Verlag, New York, Heidelberg, Berlin, 1982.

[48] J. A. Séguier, Eléments de la théorie des groupes abstraits. Paris, Gauthier-Villars, 1904.

[49] A. Shalev and E. I. Zel'manov, *Pro-p groups of finite coclass*, Math. Proc. Camb. Phil. Soc. **111** (1992), 417 – 421.

[50] A. Shalev, *The structure of finite p-groups: Effective proofs of the coclass conjectures*, Invent. Math. **115** (1994), 315 – 345.

[51] A. Shalev, *Finite p-groups*, NATO Adv. Sci. Inst. Ser. C Math. Phys. Sci. **471** (1995), 401–450.

[52] R. Shepherd, *p-groups of maximal class*, Ph.D. thesis, University of Chicago, 1971.

[53] C. C. Sims, *Enumerating p-groups*, Proc. London Math. Soc. (3), **15** (1965), 151 – 166.

[54] C. C. Sims, Computation with finitely presented groups. Cambridge University Press, 1994.

[55] O. Taussky, *Remark on the class field tower*, J. London Math. Soc. **12** (1937), 82 – 85.

[56] The GAP Group, GAP – *Groups, algorithms and programming, version 4.4*, available at www.gap-system.org.

[57] A. Vera-López, J. M. Arregi, and F. J. Vera-López, *Some bounds for the degree of commutativity of a p-group of maximal class II*, Comm. Algebra **23** no. 7 (1995), 2765 – 2795.

[58] A. Vera-López, J. M. Arregi, and F. J. Vera-López, *Some bounds for the degree of commutativity of a p-group of maximal class III*, Math. Proc. Cambridge Philos. Soc. **122** no. 2 (1997), 251 – 260.

[59] A. Vera-López, J. M. Arregi, M. A. García-Sánchez, F. J. Vera-López, and R. Esteban-Romero, *The exact bounds for the degree of commutativity of a p-group of maximal class I*, J. Algebra **256** no. 2 (2002), 375 – 401.

[60] A. Vera-López, J. M. Arregi, M. A. García-Sánchez, F. J. Vera-López, and R. Esteban-Romero, *The exact bounds for the degree of commutativity of a p-group of maximal class II*, J. Algebra **273** no. 2 (2004), 806 – 853.

[61] A. Vera-López, J. M. Arregi, and A. Jaikin-Zapirain, *On the degree of commutativity of p-groups of maximal class*, Math. Nachr. **281** no. 11 (2008), 1638 – 1650.

[62] A. Vera López and G. A. Fernández-Alcober, *On p-groups of maximal class. II*, J. Algebra no. 1 (1991), 179 – 207.

[63] A. Vera López and G. A. Fernández-Alcober, *On p-groups of maximal class. III*, Math. Proc. Cambridge Philos. Soc. **109** no. 3 (1991), 489 – 507.

[64] A. Vera-López and G. A. Fernández-Alcober, *Some bounds for the degree of commutativity of a p-group of maximal class*, Bull. Austral. Math. Soc. **51** no. 3 (1995), 353 – 367.

[65] A. Vera López and B. Larrea, *On p-groups of maximal class*, J. Algebra **137** no. 1 (1991), 77 – 116.

[66] L. R. Vermani, An elementary approach to homological algebra. Chapman & Hall/CRC Monographs and Surveys in Pure and Applied Mathematics, vol. 130, Chapman & Hall/CRC, Boca Raton, FL, 2003.

[67] A. Wiman, *Über p-Gruppen mit maximaler Klasse*, Acta Math. **88** (1952), 317–346.

List of symbols

Entries are in approximately the order in which they first occur in the text. Some entries have the given meaning in the appropriate sections of the text, but have other meanings elsewhere.

General notation: Elements and sets

p	A prime
d	The integer $p-1$
$p^\star$	The number $-(p-3)^2/4$
n	A positive integer, written as $n = \mathfrak{r}d + \mathfrak{i}$ with integers $\mathfrak{r} \geq 0$ and $1 \leq \mathfrak{i} \leq d$
$\mathbb{N}$	The set of positive integers
$\mathbb{Z}$	The set of integers
$\mathbb{Q}$	The rational numbers
$\mathbb{R}$	The real numbers
$\mathbb{C}$	The complex numbers
$\mathbb{F}_q$	The field with q elements
$\mathbb{Q}_p$	The p-adic numbers
$\mathbb{Z}_p$	The p-adic integers
θ	A primitive p-th root of unity over $\mathbb{Q}_p$
$\mathbb{Q}_p(\theta)$	The p-th local cyclotomic field defined by θ
$\mathbb{Z}_p[\theta]$	The ring of integers of the field $\mathbb{Q}_p(\theta)$
$R^\star$	The unit group of a ring R
$R[X]$	The ring of polynomials over a ring R
$\mathrm{GL}(m, R)$	The group of invertible $m \times m$ matrices over a ring R
$M(m, R)$	The R-module of $m \times m$ matrices over a ring R
$\mathcal{U}_p$	The group of units $\mathbb{Z}_p[\theta]^\star$
κ	The prime element $\kappa = \theta - 1$ of $\mathbb{Z}_p[\theta]$
$\mathfrak{p}$	The maximal ideal (κ) of $\mathbb{Z}_p[\theta]$
ω	A primitive $(p-1)$-th root of unity in $\mathbb{Z}_p$
$\mathfrak{G}$	The companion matrix of $1 + X + \ldots + X^d$ over $\mathbb{Z}_p$ with last row $(-1, \ldots, -1)$

General notation: Groups, subgroups, and invariants

$H \leq G$	The group H is a subgroup of the group G
$H < G$	The group H is a proper subgroup of the group G
$H \trianglelefteq G$	The group H is a normal subgroup of the group G
$H \triangleleft G$	The group H is a proper normal subgroup of the group G
$\langle M \rangle$	The subgroup of a group G generated by $M \subseteq G$
$H \cong G$	The groups H and G are isomorphic
$A \cong_G B$	The groups A and B are isomorphic as G-groups
$[G : H]$	The index of the subgroup H in G
G^n	The direct product $G \times \ldots \times G$ with n copies of G
$G^{[n]}$	The subgroup of the group G generated by $\{g^n \mid g \in G\}$
$\Phi(G)$	The Frattini subgroup of the group G
$\gamma_j(G)$	The j-th term of the lower central series of the group G
$\zeta(G)$	The center of the group G
$C_G(N)$	The centralizer of N in the group G

$\operatorname{Stab}_G(N)$	The stabilizer of N in the group G
$\operatorname{Aut}(G)$	The automorphism group of the group G
$C_n, C_n(x)$	The cyclic group of order n (generated by x)
$c(G)$	The nilpotency class of the group G
$cc(G)$	The coclass of the group G
$doc(G)$	The degree of commutativity of the group G

Chapter 1

D_{2^n}	The dihedral group of order 2^n	2
SD_{2^n}	The semi-dihedral group of order 2^n	2
Q_{2^n}	The quaternion group of order 2^n	2
$\mathcal{G}(p,r)$	The coclass graph associated with p-groups of coclass r	3
$\mathcal{G}_k(p,r)$	A shaved coclass graph associated with p-groups of coclass r	4

Chapter 2

$\mathcal{G}(p)$	The graph associated with p-groups of maximal class	9
S	The p-adic space group of maximal class	9
P	The point group of S	9
T	The translation subgroup of S	9
S_n	The nilpotent quotient $S/\gamma_n(S)$	9
T_n	The group $\gamma_n(S)$	9
$\mathcal{T}, \mathcal{T}(p)$	The coclass tree of $\mathcal{G}(p)$	9
$\mathcal{B}_n$	The n-th branch of $\mathcal{T}$	9
$\mathcal{D}_k(G)$	The k-step descendant tree of G	9
$\mathcal{B}_n[k]$	The shaved branch $\mathcal{B}_n \cap \mathcal{D}_k(S_n)$	9
$\mathfrak{c}$	The constant $\mathfrak{c} = \mathfrak{c}(p)$	10
$\mathfrak{e}_n$	The depth of the body $\mathcal{T}_n$	10
$\lfloor a \rfloor$	The largest integer not greater than a	10
$C_{f,e}$	The skeleton group defined by f and e	10
$\mathcal{T}_n$	The body of the branch $\mathcal{B}_n$	10
$\mathcal{S}_n$	The skeleton of the branch $\mathcal{B}_n$	10
ι, ι_n	An embedding $\mathcal{T}_n \hookrightarrow \mathcal{B}_{n+d}$	11
$\mathcal{P}(G)$	The periodicity class of the group G	12

Chapter 3

$[x,y]$	The commutator $x^{-1}y^{-1}xy$	17
$P_j, P_j(G)$	The j-th term of the refined central series of the group G	18
$G \ltimes N$	A split extension of a G-module N by G	20

Chapter 4

$\operatorname{relord}(g)$	The relative order of the element g	21
$\langle \mathcal{G} \mid \mathcal{R} \rangle$	A group presentation with generators $\mathcal{G}$ and defining relators $\mathcal{R}$	22
$u =_G w$	The words u and w represent the same element in the group G	22
$Z^i(G,N)$	The group of i-th cocycles of G with coefficients in N	25
$E(\gamma)$	The group extension defined by the 2-cocycle γ	25
$B^i(G,N)$	The group of i-th coboundaries of G with coefficients in N	25
$f\vert_M$	The restriction of the mapping f to the domain M	25
id_M	The identity mapping $M \to M$, $m \mapsto m$	25
$H^i(G,N)$	The i-th cohomology group of G with coefficients in N	25
$\operatorname{Comp}(G,N)$	The group of compatible pairs of G and N	25

List of symbols

τ	A transversal which maps a normalized word s onto $s = \tau(s)$	27
$r.x_r$	The relation $u = vx_r$ if r is the relation $u = v$	27
x_γ	The tail vector induced by the 2-cocycle γ	27
$\mathcal{E}(x)$	The presentation defined by the list x	27
$E(x)$	The group defined by the presentation $\mathcal{E}(x)$	27
$\mathcal{Z}(G, A)$	The group of tail vectors of G in A	28
$\mathcal{B}(G, A)$	The group of coboundary tail vectors of G in A	28
γ_x	The canonical 2-cocycle inducing the tail vector x	28

Chapter 5

ν_p	The p-adic valuation	31
$\|\cdot\|_p$	The p-adic absolute value	31
$\mathfrak{p}^z$	The set $\{t\kappa^z \mid t \in \mathbb{Z}_p[\theta]\}$	32
$\mathcal{G}(\mathbb{Q}_p(\theta)/\mathbb{Q}_p)$	The Galois group of the extension $\mathbb{Q}_p(\theta)/\mathbb{Q}_p$	32
$\mathcal{U}_p^{(i)}$	The group of i-th one-units	33
$\exp$	The exponential mapping	33
$\log$	The logarithm mapping	33
$\varprojlim G_\lambda$	The inverse limit of the inverse system $(G_\lambda)_\lambda$	34
P	The group $C_p(g)$	35
T_{e+1}	The group $(\mathfrak{p}^e, +)$	35
A_e	The group T/T_{e+1}	35
$T \otimes T$	The tensor product $T \otimes T$	35
$T \wedge T$	The exterior square $T \wedge T$	35
z	The element $z = \prod_{0 \leq i < j < d} \theta^i \wedge \theta^j$	37
$\widehat{z}$	A generator of $A_e^P = T_e/T_{e+1}$	37
f_k	A P-homomorphism $T \wedge T \to T$	37
$\delta_{k,j}$	The Kronecker delta	37
$\widehat{f}_k$	A P-homomorphism $T \wedge T \to A_e$	37
σ_j	The automorphism $T \to T$ defined by $\theta \mapsto \theta^{j \bmod p}$	38
F_a	A P-homomorphism $T \wedge T \to T$	38
ρ_a	A homomorphism $\mathcal{U}_p \to \mathcal{U}_p$	39
τ_a	A homomorphism $(\mathfrak{p}^2, +) \to (\mathfrak{p}^2, +)$	39
r	A primitive $(p-1)$-th root of unity in $\mathbb{F}_p$ with $r \equiv \omega \bmod p$	39
v_k	An eigenvector of σ_r with eigenvalue ω^k	40
$\omega_{a,k}$	The eigenvalue of τ_a corresponding to the eigenvector v_k	41
$p_{a,k}$	The largest integer with $\omega_{a,k} \equiv 0 \bmod p^{p_{a,k}}$	41
e_0	A bound for an integer in Theorem 5.14	42

Chapter 6

A_e	The S-module T/T_{e+1}	45
$\iota_{n,e}$	A P-module isomorphism $T_n/T_{n+e} \to A_e$	46
$S_\mathbb{Z}$	The integral version of S	46
$T_\mathbb{Z}$	The group $(\mathbb{Z}[\theta], +)$	46
$T_{\mathbb{Z},n}$	The ideal of $\mathbb{Z}[\theta]$ generated by κ^{n-1}	46
$\langle \mathcal{S} \mid \mathcal{R} \rangle$	The standard p.c.p. for the group $S_\mathbb{Z}$ with $\mathcal{S} = \{\mathfrak{g}, \mathfrak{t}_1, \ldots, \mathfrak{t}_d\}$	49
$\langle \mathcal{S} \mid \mathcal{R}_n \rangle$	The standard p.c.p. for the group S_n	49
$\langle \mathcal{A}_e \mid \mathcal{C}_e \rangle$	The standard p.c.p. for the group A_e	49
$\mathcal{A}$	The abstract generating set $\mathcal{A} = \{\mathfrak{a}_1, \ldots, \mathfrak{a}_d\}$ for A_e with $e \geq d$	49
$\mathcal{M}_e$	The conjugate relations describing the S-module structure of A_e	49
$S_{\mathbb{Z}_p}$	A group isomorphic to S	50

δ_t	The 1-cocycle $\delta_t \in Z^1(P,T)$ defined by $\delta_t(\mathfrak{g}) = t$	52
$\alpha(j,c,t)$	An automorphism of S	52
$x = (x_{i,j})$	A vector of tails $x_{i,j}$ with $0 \leq i \leq j \leq d$	53
$\mathcal{E}(x)$	The presentation defined by the vector of tails x	53
$E(x)$	The group defined by the presentation $\mathcal{E}(x)$	53
A^S	The subgroup of fixed points under the S-action on A	54
α_k	The element $\alpha_k = \mathfrak{g} + \ldots + \mathfrak{g}^{p-k}$ in $\mathbb{Z}_p P$	54
β_k	The element $\beta_k = 1 + \mathfrak{g} + \ldots + \mathfrak{g}^{d-k}$ in $\mathbb{Z}_p P$	54
Γ_f	The 2-cocycle defined by the P-homomorphism f	56
x_f	The tail vector x_{Γ_f} defined by f	56
$\mathrm{H}(n,e)$	The group of hom tail vectors	56
$\mathrm{M}(n,e)$	The group of mainline tail vectors	57
$\mathrm{T}(n,e)$	The group of twig tail vectors	57
$\widehat{\mathrm{H}}(n,e)$	The hom tail vectors induced by surjective homomorphisms	58
$\mathrm{L}(n,e)$	The hom tail vectors induced by liftable homomorphisms	58
$\widehat{\mathrm{L}}(n,e)$	The intersection of $\widehat{\mathrm{H}}(n,e)$ and $\mathrm{L}(n,e)$	58
$\mathrm{TH}(n,e)$	The direct sum $\mathrm{T}(n,e) \oplus \mathrm{H}(n,e)$	58
$\widehat{\mathrm{TH}}(n,e)$	The direct sum $\mathrm{T}(n,e) \oplus \widehat{\mathrm{H}}(n,e)$ if $e \geq 2$	58
$\mathrm{H}_t(n,e)$	A complement to $\mathrm{L}(n,e)$ in $\mathrm{H}(n,e)$	59

Chapter 7

$\gamma_{n,e}$	The canonical 2-cocycle in $Z^2(S_n, T_n/T_{n+e})$ defining S_{n+e}	62
$\gamma'_{n,e}$	The canonical 2-cocycle in $Z^2(S_n, A_e)$ defining S_{n+e}	62
$\mathfrak{m}_{n,e}$	The mainline tail vector of S_n in A_e defining S_{n+e}	62
$\mathrm{Comp}(e)$	The group of compatible pairs of S and A_e	65
$\Sigma_{n,e}$	The stabilizer $\mathrm{Stab}_{\mathrm{Comp}(e)}(\mathfrak{m}_{n,e})$	66
$\xi_{n,e}$	A homomorphism $\mathrm{Aut}(S) \to \mathrm{Aut}(A_e)$	66

Chapter 8

$\nu_{n,e}$	An isomorphism $\mathcal{Z}(n,e) \to \mathcal{Z}(n+d,e)$	71

Chapter 9

$\mathcal{D}_k(G)$	The k-step descendant tree of the group G	75
$\Pi(G)$	A periodic parent of the group G	75
$x_{f,e}$	The hom tail vector $x_{\pi \circ f}$ where $\pi \colon T \to A_e$ and $f \colon T \wedge T \to T$	76
$E_{m,e}(f)$	The group $E(\mathfrak{m}_{m,e} + x_{f,e})$ defined by f	76
$\Sigma_{m,e}(f)$	The stabilizer $\mathrm{Stab}_{\mathrm{Aut}(S)}(x_{f,e})$	76
$\mathcal{Z}_l(m,e,k)$	A group of tail vectors of S_m in T_{e+1}/T_{e+1+k}	76
$\mathcal{Z}_t(m,e,k)$	A group of twig tail vectors of S_m in T_{e+1}/T_{e+1+k}	76
$\mathcal{Z}(m,e,k)$	The group $\mathcal{Z}_l(m,e,k) \oplus \mathcal{Z}_t(m,e,k)$	76
H_e	The group $\mathrm{Hom}_P(T \wedge T, T_{e+1})$	77
$\phi_{m,j,k}$	An isomorphism $\mathcal{Z}(m,j,k) \to \mathcal{Z}(m,j+d,k)$	79
$\phi^l_{m,j,k}$	The restriction of $\phi_{m,j,k}$ to $\mathcal{Z}_l(m,j,k)$	79
$\phi^t_{m,j,k}$	The restriction of $\phi_{m,j,k}$ to $\mathcal{Z}_t(m,j,k)$	79
$\widehat{\Sigma}_{m,e}(f)$	The stabilizer $\mathrm{Stab}_{\mathrm{Aut}(S_{m+e})}(x_{f,e})$	80

Index

0–9

1-coboundary 26
1-cocycle 26
2-coboundary 25
2-cocycle 25
2-step centralizer 18

B

body of a branch 10
branch of a coclass tree 4, 9

C

consistent p.c.p. 22
canonical 2-cocycle 28
canonical transversal 50
capable 9
central extension 20
coboundary tail vector 28
coclass 2
    coclass conjectures 2, 95
    coclass graph 3
    coclass tree 4, 9
    shaved coclass graph 4
cohomology group
    first cohomology group 26
    second cohomology group 25
collection 23
commutator 17
compatible pairs 25, 64
component
    hom component 58
    mainline component 58
    twig component 58
Conjecture A 3, 95
Conjecture D 4, 95
Conjecture P 5
conjugate relations 22
consistency checks 24
consistent polycyclic presentation . . . 22
constructible group 10

D

degree of commutativity 18
depth
    of a tree 4, 10
    of a vertex 4, 10
descendant 3, 9
    k-step descendant 76
descendant tree 4, 9
    k-step descendant tree 9
dimension of space group 10
directed set 34

E

equivalent word 22
exact sequence 55
exponential mapping 33
extension 20, 24
    central extension 20
    split extension 25
exterior square 35

F

Frattini subgroup 17
free group 22

G

G-group 19
G-module 19
Galois complement 81
group presentation 22
group extension 24
group of i-th one-units 33

H

Hensels Lemma 31
hom component 58
hom tail vector 56

I

i-th one units 33
immediate descendant 3, 9
inflation homomorphism 55
internal direct product 33

K

Kronecker delta 37

107

L

liftable homomorphism 37, 58
logarithm mapping 33
lower central series 17

M

mainline 4, 9
    mainline component 58
    mainline group 9
    mainline tail vector 57, 62
maximal class 1

N

nilpotency class 1
normal form 21
normalized word 22

O

one-units 33
operator group 34

P

p-adic
    p-adic units 31
    p-adic absolute value 31
    p-adic integers 31
    p-adic number field 32
    p-adic numbers 31
    p-adic valuation 31
p-group 1
p'-part 14
p-th local cyclotomic field 32
p.c.p. 22
parameterized presentation 2, 50
parent 3, 9
    d-step parent 14
periodic parent 13, 75
periodicity
    periodicity class 3, 12, 73
    periodicity class of type 2 88
    periodicity mapping 11
    periodicity of type 1 . . . . 11, 72
    periodicity of type 2 . . . . 13, 75
point group 9, 45
polycyclic
    consistent polycyclic presentation 22
    polycyclic presentation 22
    polycyclic sequence 21
    polycyclic group 21
    polycyclic series 21
power relations 22
presentation 22
prime element 32

pro-p group 4

R

rank (of a free module) 33
refined central series 18
relative order 21
residue class field 32
restriction homomorphism 55

S

shaved
    coclass graph 4
    branch 9
    subtree 9
skeleton
    skeleton group 10, 69
    skeleton of a branch 10
space group of maximal class 45
    integral version 46
split extension 20, 25
standard presentation 49
strongly isomorphic 25

T

tail vector 27, 28, 53
    coboundary tail vector 28
    hom tail vector 56
    mainline tail vector 57, 62
    tail vector induced by a cocycle 27
    tails 27
    twig tail vector 57
tensor product 35
terminal 9
topological group 34
transgression homomorphism 55
translation subgroup 9, 45
transversal 24
twig
    twig component 58
    twig group 10
    twig tail vector 57
    twigs 10

U

uniserial action 19

V

valuation ring 31, 32

W

width of a tree 4, 10
word in $\mathcal{G}$ 21

Acknowledgments

First of all, I would like to thank my supervisor, Professor Dr. Bettina Eick, for the guidance and very valuable support throughout my years as a diploma and doctoral student. She was always kindly willing to listen to my problems and I am deeply grateful for her outstanding efforts which have made this thesis possible. I also thank her for encouraging me to take part in academic life and for her support of my numerous conference attendances. It was always a pleasure to work under her supervision.

I would like to thank all my friends and colleagues at the TU Braunschweig for the great time I have spent since the beginning of my study in autumn 2000. I am especially thankful to Professor Dr. Heiko Harborth who convinced me to study mathematics when I was a school student, and who supported me continuously throughout my study.

I would like to thank Professor Charles R. Leedham-Green, Professor Eamonn O'Brien, and Professor Marcus du Sautoy for comments and helpful discussions in the course of preparing this thesis. Moreover, I am particularly indebted to Tobias Roßmann for proofreading and constructive criticism. His comments and suggestions were always of great value. I also thank Dörte Feichtenschlager, PD Dr. Harm Pralle, Christian Sievers, and Andreas Wörner. They have all been a great help to me.

Warmest thanks go to my parents, family, friends, and, especially, to my wife Stefanie, for their overwhelming support and patience during the last years.

Last but not least, I had the great honor to become a scholarship holder of the *Studienstiftung des deutschen Volkes* and I am grateful for the financial support and all the encouragement beyond study. I also want to thank the *Braunschweigischer Hochschulbund* for several travel grants.

Die VDM Verlagsservicegesellschaft sucht für wissenschaftliche Verlage abgeschlossene und herausragende

Dissertationen, Habilitationen, Diplomarbeiten, Master Theses, Magisterarbeiten usw.

für die kostenlose Publikation als Fachbuch.

Sie verfügen über eine Arbeit, die hohen inhaltlichen und formalen Ansprüchen genügt, und haben Interesse an einer honorarvergüteten Publikation?

Dann senden Sie bitte erste Informationen über sich und Ihre Arbeit per Email an *info@vdm-vsg.de*.

Sie erhalten kurzfristig unser Feedback!

VDM Verlagsservicegesellschaft mbH
Dudweiler Landstr. 99   Telefon +49 681 3720 174
D - 66123 Saarbrücken   Fax   +49 681 3720 1749
www.vdm-vsg.de

Die VDM Verlagsservicegesellschaft mbH vertritt

Printed by Books on Demand GmbH, Norderstedt / Germany